Back Row from Left: Professor Vyacheslav S. Yastrebov, Dr. Marvin K. Moss, Dr. Craig E. Dorman, Dr. Pierre Papon

Front Row from Left: Professor Zeshi Chen, Mr. Stephan B. MacPhee, Mr. Isao Uchida, Dr. Joseph T. Baker, Dr. Tomio Asai

International Symposium on
New Directions of Oceanographic R&D

N. Nasu · S. Honjo (Eds.)

New Directions of Oceanographic Research and Development

With 36 Figures

Springer-Verlag
Tokyo Berlin Heidelberg
New York London Paris
Hong Kong Barcelona
Budapest

Dr. NORIYUKI NASU
Professor Emeritus, University of Tokyo
Professor, University of the Air, 2–11, Wakaba, Mihama-ku, Chiba, 261 Japan

Dr. SUSUMU HONJO
Department of Geology and Geophysics
Woods Hole Oceanographic Institution, MA 02543, USA

ISBN-13:978-4-431-68227-1   e-ISBN-13:978-4-431-68225-7
DOI: 10.1007/978-4-431-68225-7

Printed on acid-free paper

Library of Congress Cataloging-in-Publication Data
New directions of oceanographic research and development/N. Nasu, S. Honjo (eds.). p. cm.
Includes bibliographical references and index. Based on the International Symposium on New
Directions of Oceanographic Research and Development, held Nov. 20–21, 1991 in Tokyo under
the sponsorship of the Japan Marine Science and Technology Center. ISBN-13:978-4-431-68227-1
1. Oceanography—Research—Congresses. I. Nasu, Noriyuki. II. Honjo, Susumu. GC57.N49, 1993.
551.46'0072—dc20. 92-32507

© Springer-Verlag Tokyo 1993
Softcover reprint of the hardcover 1st edition 1993

Preface

The earth where we live is the only planet of our solar system that holds a mass of water we know as the ocean, covering 70.8% of the earth's surface with a mean depth of 3,800 m. When using the term ocean, we mean not only the water and what it contains, but also the bottom that supports the water mass above and the atmosphere on the sea surface. Modern oceanography thus deals with the water, the bottom of the ocean, and the air thereon. In addition, varied interactions take place between the ocean and the land so that such interface areas are also extended domains of oceanography.

In ancient times our ancestors took an interest in nearshore seas, making them an object of constant study. Deep seas, on the other hand, largely remained an area beyond their reach. Modern academic research on deep seas is said to have been started by the first round-the-world voyage of Her Majesty's R/V Challenger I from 1872 to 1876. It has been only 120 years since the British ship left Portsmouth on this voyage, so oceanography can thus be considered still a young science on its way to full maturity.

Despite the short historical background, academic accomplishment in oceanography has indeed been remarkable, and its contributions have led to great exploration and utilization of the ocean. Since the Second World War, the climate for international cooperation in the field of oceanography has become favorable, and studies have accelerated not only on a local or regional basis but also on a global basis, with countries being involved bilaterally and multilaterally.

Remarkable advances in modern technology have had the fortunate effect of causing dramatic improvements and scale enlargements in oceanographic research. Demands for more sophisticated oceanographic research and work at greater depths have been raised by those engaged in ocean exploration and exploitation, and this has also accelerated academic development of oceanography. The need for submarine petroleum and gas exploitation on continental shelves, for example, has prompted oceanographers to develop an understanding of the subsurface structure of the shelf and the behavior of the seawater moving above it. Work for harbor construction, land reclamation, protection of beaches from erosion, and coastal mariculture have also required precision oceanic studies of near-shore processes, including certain aspects of biodiversity.

By nature, the study of the vast ocean calls for its understanding as one whole body. Accumulation of local, well-balanced data becomes essential to understanding the real nature of the ocean through global scale studies. More recently, preserving the earth's environment and preventing its degradation on a global basis have become matters of serious public concern. In terms of scientific knowledge, however, we stand far behind where we should be. For example, we know very little about eustatic sea level changes in relation to the fluctuations of carbon dioxide in the atmosphere. A new mission has now been added to oceanographic research, which, as observed above, has been motivated mainly by academic interest and by the dictates of ocean exploration and exploitation. This mission, calls upon us to join forces to provide a sound scientific base for "sustainable development", without which human beings and other forms of life on earth would face grave difficulty.

The circumstances being as stated above, it is an appropriate period in which to hold a symposium aiming to exchange the latest information on oceanographic studies conducted by the leading research groups of the world. The International Symposium on New Directions of Oceanographic R&D, November 20–21, 1991, Tokyo, held under the sponsorship of the Japan Marine Science and Technology Center (JAMSTEC), was honored with the presence of directors from several leading oceanographic institutions around the world. Each director discussed the activities of their institution and left a strong impression as to the necessity of international ties. The special lectures and panel discussions were also beneficial in that they were inspiring and thought-provoking.

We sincerely hope that the reader of this book can feel the new currents in international cooperation between oceanographic research institutions around the world. Finally, it is our great honor to have the opportunity to write a preface to this volume and we would like to extend our sincere gratitude to all of those who participated in the symposium and gave us many new and valuable perspectives.

NORIYUKI NASU
SUSUMU HONJO

Welcome Messages

The Japan Marine Science and Technology Center is celebrating its 20th anniversary this year. As a core research institute of marine science and technology in Japan we have been able to produce research results in various aspects of marine development with the cooperation of industry, academia, and government. The interest in the ocean among the general public has never been as high as it is today, because of concerns with global-scale environmental change, for example. The more we recognize the importance of the role that the ocean plays, the more we come to realize how little we know about the ocean, such as its influence on the climate. It is a common task for all mankind to understand the ocean occupying 70% of the earth's surface well and to utilize it wisely. To undertake this task, it is essential that all countries of the world cooperate in solving the problems before us. JAMSTEC is prepared to contribute what it can to the solution of the problems and thus to the world.

For the leaders of major research institutes of the world, it is meaningful to get together, listen to what others are doing and planning to do, and exchange views in order to identify future directions of research and development. This is why we decided to organize this symposium. This symposium gives us the opportunity to renew our pledge to do our utmost not only for the good of the international community but also for responding to the needs of domestic sectors. So, representing the entire JAMSTEC staff, I would like to ask for your continued cooperation and support for our institute. Thank you.

KOSAKU INABA

Thank you very much for the introduction. Because of the cabinet reshuffle which took place recently, I have now assumed the post of Minister of State for Science and Technology.

It is very timely to have this international symposium on the future directions of oceanographic research and development with the presence of leaders from major oceanographic research institutes of the world on the occasion of the 20th anniversary of JAMSTEC, when there has come into being strong recognition of the need to develop and utilize the ocean and when the ocean has come into the spotlight internally as well as internationally because of global environmental concerns. I would like to say a few words of congratulation.

The ocean is a treasure-house of resources vital to the survival of mankind. At the same time, it is the final frontier left on the Earth for man. We need to make every effort possible to advance research on the ocean to bring about development which is compatible with the preservation of the environment.

In order to effectively cope with the common issues of mankind, such as global warming and deterioration of the environment, we will have to have a much better understanding of the ocean, which has a strong influence on the environmental changes of the Earth. I really hope that this symposium will help toward that end. Japan is an island country; we have been deeply involved with the ocean since olden times. In recent years, Japanese science and technology has, in the main, reached world levels. From now on, we should endeavor to enhance basic research from the global perspective and contribute actively to *the world in ocean science*. From this point of view, we have been participating in a number of joint international oceanographic and research programs that are currently underway or are planned. At the same time, it is important for us to strive to develop new, or to improve existing, high technology systems and facilities, such as Shinkai 6500 which was developed by JAMSTEC and is now commemorating its 30th anniversary.

In June next year, there will be a United Nations conference on environment and development to discuss ways of attacking issues regarding the Earth's environment on a global scale. We are also about to start a global observation network of the ocean which covers 70% of the Earth's surface. So it is meaningful that we can organize this international symposium at this time, I would like to thank all of you who have come far to participate in this event. I would also like to pay my respects to the staff of JAMSTEC who have made this symposium come true. I really hope that there will be very active and useful discussions today and tomorrow on the future directions of oceanic research and development. I sincerely hope this will also contribute to the promotion of international cooperation in this area for the immediate future as well as for the formulation of long-term prespectives.

KANZO TANIGAWA

Contents

Oceanographic Institutions of the World-Current Activity and Future Plan

Special Lecture

Panel Discussions

Summary and Closing Address

List of Contributors

Asai, T.   96
Baker, J.   3, 186
Baker, J.T.   186
Chen, Z.   44, 186
Dorman, C.E.   111, 159
Frieman, E.A.   122
Honjo, S.   186
Hotta, H.   159
Ishii, S.   186
Karube, I.   159
Kobayashi, K.   159
Kondo, J.   143

Laubier, L.   159
MacPhee, S.B.   15
Moss, M.K.   186
Nakanishi, T.   186
Nakato, H.   159
Nasu, N.   159
Nozaki, Y.   186
Papon, P.   60
Ross, D.I.   186
Uchida, I.   78, 219
Yastrebov, V.S.   138, 159

Oceanographic Institutions of the World — Current Activity and Future Plan

Australia

Interdisciplinary Oceanographic Studies at the Australian Institute of Marine Science

Joe Baker[1]

Key words. Interdisciplinary — Oceanographic — Marine science — Australian — Coastal — Processes — Reef — Environmental — Biotechnology — Tropical

Summary. The Australian Institute of Marine Science (AIMS) concentrates its oceanographic research activities on the tropical coastal zones and on coral reefs, with consideration of both open ocean and land-based activities which impact on these marine ecosystems. In this way the activities of the AIMS relate directly to those areas of major concern with respect to ecologically sustainable development in tropical coastal waters, e.g., out to the edge of the continental shelf. Research and development activities are planned and operated in 5-year rolling programs, with the current operational period being from July 1, 1991 to June 30, 1996. Active national and international collaboration is sought and encouraged.

This paper outlines the philosophy of operation of AIMS and the way in which strategic R&D planning can accommodate the requirements of maintaining excellent fundamental research efforts, while applying the results of that research to community, industry, and government requirements. The paper illustrates how the R&D operations are concentrated within four major programs: Coastal Processes and Resources, Coral Reef Ecosystems, Environmental Studies and Biotechnology, and Tropical Oceanography. These four programs are closely integrated, interdisciplinary efforts with the principal objective of understanding marine ecosystems and processes, as well as facilitating the development of products from the marine ecosystems. The paper also addresses the challenge of long-term understanding and management of Australia's Exclusive Economic Zone.

[1] Australian Institute of Marine Science, PMB. 3. M.C., Townsville, Queensland, 4810, Australia

Establishment of the Australian Institute of Marine Science

The Australian Institute of Marine Science (AIMS) is an agency with principal funding derived from the Australian Government and established by an act of Parliament in 1972. The Institute is governed by a Council appointed by the Australian Government and is located on 207 ha of land excised from National Park at Cape Ferguson on Cape Cleveland approximately 50 km south of Townsville, North Queensland. It is adjacent to the Great Barrier Reef and exists on part of the Australian coastline characterised by mangroves, coastal estuaries, and rich fringing and outer coral reefs. The Institute began operating in 1974 in temporary buildings near Townsville and the relocation to the existing site was effected in 1977. The site provides direct access to a variety of marine habitats which are part of the Great Barrier Reef Marine Park and the near offshore waters are declared a Scientific Research Zone within the Great Barrier Reef Marine Park.

The Functions of the Australian Institute of Marine Science

The Australian Institute of Marine Science Act 1972 sets out the function of the Institute as follows:

1. To carry out research in marine science
2. To arrange for the carrying out of research in marine science by any other institution or person
3. To cooperate with other institutions and persons in carrying out research in marine science
4. To provide any other institution or person with facilities for carrying out research in marine science or otherwise assist any other institution or person in carrying out research in marine science
5. To collect and disseminate information relating to marine science and, in particular, to publish reports, periodicals, and other papers relating to marine science
6. To do anything incidental or conducive to the performance of any of the foregoing functions

Principal Geographic Areas of Study

The main geographic area of interest of all AIMS research is presently directed to the coastal and continental shelf regions of tropical Australia. The research programs are essentially directed to achieving an understanding of the marine environment in this area. The geographic restriction reflects AIMS' financial, technical, and operational capabilities (particularly shipping) and its understand-

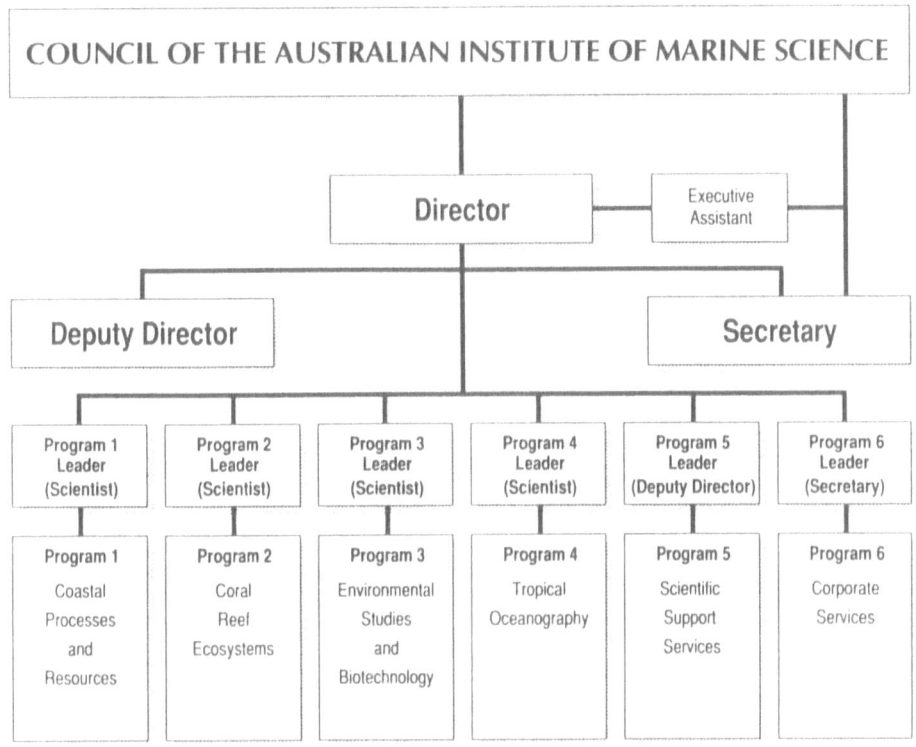

Fig. 1. Organizational Structure of the Australian Institute of Marine Science (AIMS)

ing with other national agencies (CSIRO, GBRMPA, BMR) and universities which work in oceans surrounding Australia.

In certain circumstances consideration is given to projects which involve deep or blue water oceanography. In such cases, the specific relationships and connections between the deep seas and continental shelf, and their importance to the continental shelf processes must be demonstrated. Any deep water (greater than 300m) projects must be within the technical, operational and financial capability of AIMS or be conducted in collaboration with other agencies.

While the principal research emphasis is on the Australian marine environment, consideration is given to projects, in other regions and other countries, which will build on the expertise already developed at AIMS and lead to a better understanding of different types of tropical marine environments.

Organisation, Interactions, and Forward Planning

Figure 1 shows the organisation of the Institute and the division into programs which represent four research programs and two support programs. The four research Programs will be dealt with in detail at a later stage.

The support programs, Scientific Support Services and Corporate Services, represent one of the unique strengths of the Institute in Australia, in that the level of expertise in these areas — particularly in the Scientific Support area — are at a higher level than that available in any other marine institution in Australia. Special equipment is fabricated in the Institute and there is very close interaction between the research scientists and the skilled people in all areas of Scientific Support Services. Corporate Services also provides essential high quality services to ensure that the operations of the Institute are consistent with Government guidelines and also to facilitate the commercial interactions that the Institute must develop by reason of recent Government initiatives.

The Australian Institute of Marine Science has a wide range of interactions and is both interdisciplinary and interinstitutional in its operations. For example, in the AIMS management structure, there are no traditional departments representing disciplinary areas. It has been our experience that the artificial demarcations developed by humans do not assist in the understanding of complex ecosystems and it is essential that experts in different disciplines work together in order to better understand marine processes and marine systems.

At present the Institute has interactions with colleagues in 88 different institutions in 25 different countries in the world. Our forward planning allows for the development of 5-year rolling programs and we have as from July 1, 1991 begun research and development activities within projected (or planned) R&D activities for the quinquennium 1991–1996. This document is being published and distributed widely in order to encourage collaboration in Australia and all other countries.

A recent decision of the Government is that AIMS should move over a 5-year period to achieve a level of applied research representing 25% of its total effort. Additionally AIMS should move to achieve external earnings of 30% of the amount provided by Government, thus demonstrating the value of the research results to industry and commerce. The scope of our interactions is demonstrated in Fig. 2.

The nature of AIMS research, and the results deriving from it, is such that we can often help the Government of our country and the Governments of other countries by saving money rather than by making money, and our knowledge on the biological productivity of different coastal features, such as mangroves and seagrass beds, as well as the physical protection they afford, should be taken into account in all planning processes for coastal development in tropical regions. Fundamentally our work is concentrated on two major ecosystem types: mangroves and coral reefs. In the study of those ecosystems we investigate the waters and sediments that move within, between, and among the ecosystems; we investigate the products that can be derived from those ecosystems, whether the products be food or fine chemicals or pharmaceuticals; and we are also interested in interpreting the secrets that are locked within the different species which inhabit mangroves and coral reefs and/or the skeletons of those species.

The concentration of effort by AIMS on the continental shelf waters of the tropical regions of Australia allows the Institute to address areas of study of great

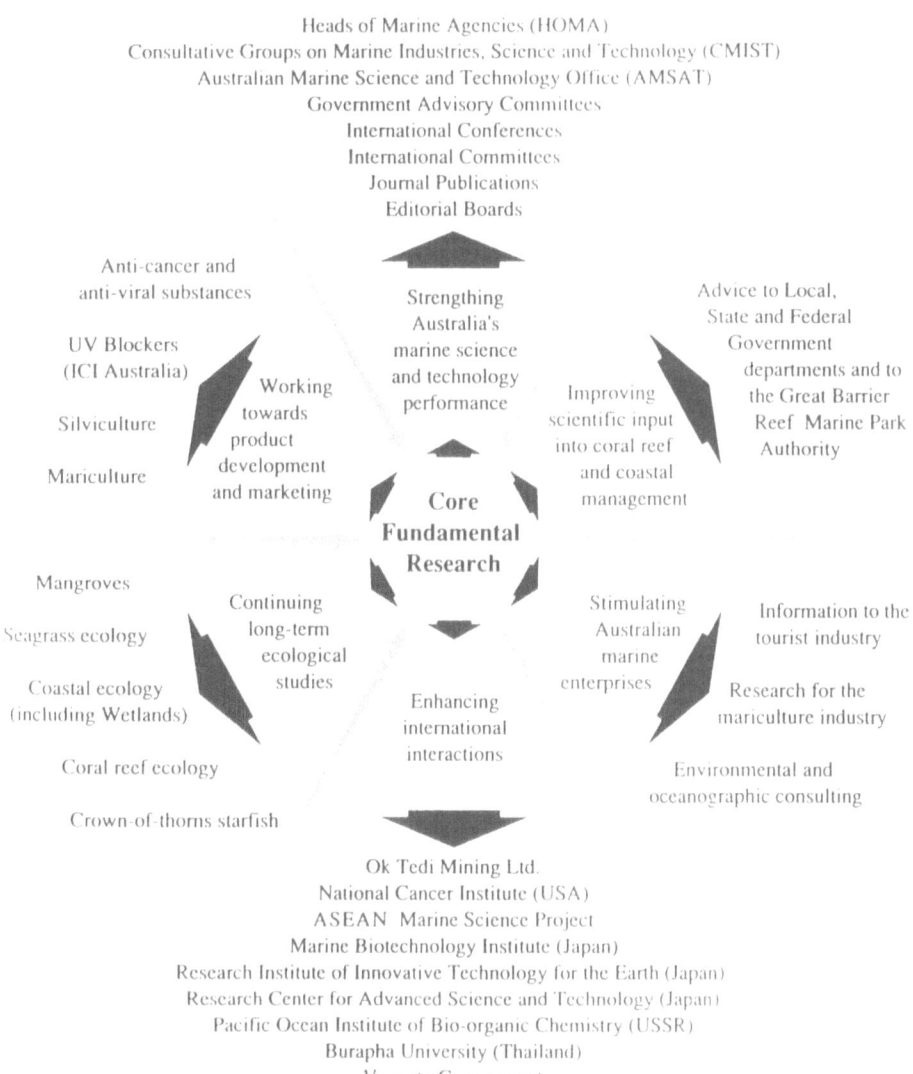

Fig. 2. Interaction chart demonstrating how the Institute's work extends from core fundamental research into product development and interactions with Government and industry

intellectual challenge. Additionally, the results of the research can be taken into account in the wise planning of development of countries within the tropical regions and also to assist those countries which have carried out development at a stage when scientific knowledge was inadequate to know the consequences of disastrous human actions, in order to move to correct the resulting degradation. In this way AIMS has a significant role to play in long-term programs understanding ecological processes related to environmental quality of tropical eco-systems. The challenges are not as romantic as those of the deep waters but they are as great. The diversity of processes in complex areas such as coral reefs and mangroves is unmatched anywhere else in the world.

Principal Research Areas

- Program 1 — Coastal Processes and Resources
- Program 2 — Coral Reef Ecosystems
- Program 3 — Environmental Studies and Biotechnology
- Program 4 — Tropical Oceanography

These research projects will be dealt with sequentially. They are undertaken within the general mission of the Institute, i.e., "To undertake research and development, to generate new knowledge in marine science and technology, promote its application in industry government and ecosystem management, and undertake complementary activities to disseminate knowledge, collaborate effec-tively, assist in the development of a National Marine Science Policy and enhance the Institute's standing as a centre of excellence."

Budget

The budget of the Institute is approximately ¥1.7 billion per year with approxi-mately ¥0.33 billion being provided as appropriation from the Federal Govern-ment of Australia and A\$3.3 million being earned from non-appropriation sources.

Staffing

The staffing of the Institute maintained by direct funding from the Government is approximately 114. Several key areas of operation are contracted to expert groups and additional staff are recruited from funding from external sources. At present the total number of staff working at the Institute is approximately 192.

Details on the Different Research Projects

Program 1—Coastal Processes and Resources

Objective

To conduct and collaborate in research into the function of tropical coastal ecosystems in order to develop an integrated understanding of the coastal marine environment for the advancement of scientific knowledge and the conservation and sustained-yield management of these natural resources.

Goals

- To gain an understanding of the ecology of mangrove tree species, particularly their responses to soil and climatic factors and their use for plantation and reclamation planting
- To describe ecological processes within coastal habitats and identify the factors controlling the structure and function of coastal and shelf ecosystems
- To investigate the transport and processing of sediments and dissolved nutrients in the coastal zone
- To provide technical advice on methods of coastal assessment to Australian and ASEAN managers

In response to the needs of the human community for sustained-yield development, a major new thrust for this project will be investigations of the potential for some mangroves as plantation species, targeting the cabinet timber market. Mangrove timber is already imported to Australia. The project will investigate silvicultural practices and sediment conditions required for maximum timber yield, as well as tissue culture techniques required for propagation of seedlings. In addition, since removal of mangrove forests is occurring on a large scale in Australia and elsewhere in the world this project will also investigate the methods of mangrove forest restoration.

Decaying wood has been shown to be a key element in mangrove food webs, and research has shown that there is no net flux of dissolved nutrients from mangrove forests. However, several areas of research require attention. For instance, there is little knowledge about the role of the rhizosphere in mangrove forest growth and productivity. Hence research in mangrove swamps in the next 5-year period will concentrate on the turnover and transfer of nutrients within the sediment/root complexes of mangrove forests. This work will be set within the focus of the project investigations of nutrient inputs to coastal systems.

The Coastal Trophodynamics Project will widen its focus during the next 5 years to include studies on other important coastal habitats and to concentrate on connectivity between habitats. In particular, coastal seagrass beds, which are crucial nursery habitats for juveniles of commercially harvested prawns, will be the focus of considerable attention as to the influence of sediment and nutrient inputs on seagrass system function. Other habitats which will receive detailed study include freshwater wetlands and coastal reef systems. Freshwater wetlands, saltmarshes, and mangrove forests form a complex of interacting habitats along much of the tropical coast. Fish such as barramundi are dependent on different parts of this mosaic during phases of their life cycle. The project will investigate the ecological connections between these habitats. Coastal reefs, as well as shallow-water seagrass beds, often bear the brunt of coastal development. During the next 5 years the program will investigate the ways in which ecological processes on coastal reefs are constrained by terrestrially derived inputs.

The program will also be involved in the monitoring of dissolved nutrients in a variety of river systems in North Queensland. Combined with runoff measurements, these data will allow estimates to be made of the influence of riverine inputs to shallow coastal shelf systems.

An integral part of all of the above biological studies is our ability to predict the movement of water and suspended sediments in coastal systems. During the next 5 years studies in this vital area of research will continue, focusing on estuarine-shelf connections in northern Australia, Papua New Guinea, Thailand, and East Africa.

The fate of transported sediments on the continental shelf will also be a focus of research. During the previous 5 years we began investigations of the shelf benthos, and now have some preliminary understanding of the way in which inputs of terrestrial sediment influence benthic food chains in some shallow shelf habitats. The aim of research in the next quinquennium is to further this work, particular emphasis being given to food chains supporting commercially harvested prawns in northern Australia and Papua New Guinea.

Program 2—Coral Reef Ecosystems

Objective

To strengthen knowledge and understanding of coral reef ecosystems and to provide a sound basis for their future conservation, development, and management.

Goals

- To increase skills in interpretation and prediction of spatial and temporal variability in reef communities and processes
- To interpret the significance of purported degradation of coral reefs in the context of inherent variability of reefs in time and space
- To provide soundly based scientific advice on appropriate strategies for conservation, development, and management of reefs and reef resources over a wide range of spatial scales

This program sees anthropogenic disturbance and exploitation of coral reefs as phenomena which must be understood within the context of their normal spatial and temporal variability in structure, diversity, and productivity. The limits within which reefs may vary must be defined and understood, and research projects have been developed accordingly.

The scope of the program is primarily biological, and includes studies which focus on particular plant or animal groups (notably corals, algae, sponges, fish) and those which emphasize ecosystem processes which cross taxonomic groupings (notably trophodynamic studies). Included are studies of evolution and of systematics (corals and fish), taphonomy (corals and crown-of-thorns starfish), physiology (corals, algae, and sponges), demography, and life histories (corals, algae and fish). Among those which focus on processes are studies of food web dynamics of the benthos/fish/plankton subsystem and benthic community response to disturbance (physical, biological, and nutritional, both natural and anthropogenic). A major multidisciplinary project on causes and consequences of outbreaks of crown-of-thorns starfish includes seven individual studies. Many projects involve original invention or refinement of existing procedures and technology for measuring, sampling and analysis.

The program's research addresses many issues of primary importance to conservation, development, and management well into the twenty-first century. An expert system based on program results and expertise will be investigated for its potential in these areas.

Studies of the major sessile benthic organisms (corals, algae, sponges) will provide an improved basis for interpreting current reef status and for predicting the likely consequences of environmental change, including greenhouse-related sea level rise, on various categories of reefs (e.g., coastal or offshore, polluted or unpolluted). Fish research will provide a better ecological basis for understanding the effects of fishing on reefs and, in particular, for distinguishing natural variability from anthropogenic effects. Trophodynamic and bio-accumulation research will provide a better basis for understanding likely consequences of overfishing on reef communities and of nutrient and pollutant pathways on coral reefs.

Program 3—Environmental Studies and Biotechnology

Objectives

- To develop an understanding of the manner in which selected marine organisms respond to and record environmental conditions
- To develop techniques to measure and assess these responses and, hence, generate proxy climate records which will contribute to understanding of the variability of climatic and other environmental processes
- To develop an understanding of the manner in which biological solutions can be applied to human requirements
- To develop an understanding of the genetics and associated environmental aspects of selected marine organisms related to their commercial mariculture

Goals

- To gain an appropriate understanding of the ways in which massive corals respond to and record their environment, and to develop methodologies and techniques to quantify these responses to produce proxy records of past environmental conditions
- To develop statistically sound, reliable reconstructions of past environmental conditions, to make use of such data, and to make it available to other users
- To gain an appropriate understanding of the physiological and biochemical responses of marine organisms to exposure to solar visible and ultraviolet (UV) radiation
- To isolate substances from marine organisms for the purposes of evaluation of their potential application for human benefit (e.g. antiviral, antitumor

substances) and evaluation for potential therapeutic value in other human diseases

- To understand the development, genetics, and environmental aspects of the mariculture of penaeid prawns

This program seeks to understand and to quantify the manner in which selected marine organisms respond to environmental conditions. Emphasis is placed specifically on responses which yield insights that are of value in satisfying human needs. Program research is now concentrated in three areas of interest: environmental records from the skeletons of reef-building corals, marine natural products chemistry, and mariculture.

Massive reef-building coral skeletons are an important, potential source of proxy environmental records. One area of proposed research focuses on the nature and causes of annual density banding in the skeletons and an understanding of how environmental parameters control or moderate the mechanisms causing coral density bands. Proxy environmental records, several 100 years long, would be an important asset for climatic modellers and would make an important contribution to the ongoing international debate about global climate change.

Coral skeletons also incorporate natural tracers which may serve as proxy signatures of past environments. For example, freshwater input to the inshore region may be reconstructed by examining fluorescent sequences in cores from massive coral colonies. This information is of immediate relevance to hydrology and climatology. In a series of national and international collaborations, isotopic signatures from coral cores are also being used to investigate past environments.

Shallow-water marine organisms receive large doses of solar UV radiation, including the potentially damaging UV-B. Many organisms respond by synthesizing UV-B radiation-absorbing compounds. During the previous quinquennium, one particular class of these compounds served as a model for the synthesis of commercial analogues for use in personal sun care products. Commercialization of these analogues continues through an established industrial research and development agreement. During the next quinquennium, additional natural UV-radiation-absorbing compounds will be examined. The projected physiological and biochemical studies are not only of scientific but also of commercial relevance.

Benthic marine organisms are a proven source for biologically active compounds. In the next quinquennium these organisms will continue to be collected as potential sources for compounds that have antitumour, antiviral, antifungal, or immunomodulatory activity. The isolation of biologically active compounds will be conducted both at AIMS and through a variety of national and international collaborations. The culture of species that have relevance as a source of interesting compounds will be investigated.

Mariculture is a developing facet of Australian primary industry. AIMS research in support of this development focuses on genetic research relevant to penaeid prawn mariculture. Research into reproductive physiology will seek to "close the life cycle" so as to enable the maintenance of controlled stocks of breeding prawns. The genetic variation in wild and cultured populations will be assessed. Additional research will assess the extent to which key traits of commercial importance, such as growth rate, are under genetic control.

Program 4—Tropical Oceanography

Objective

To develop a quantitative understanding of physical, chemical, and biological phenomena, processes, and dynamics in tropical marine waters, with an emphasis on the shelf and oceanic systems bordering northern Australia.

Goals

- To develop numerical analyses, quantitative budgets, and numerical simulation models of oceanographic and ecosystem processes in the marine tropics
- Through measurement, analysis, and modelling, define key events, communities, and processes which regulate the structure, productivity, connectivity, and dynamics of northern Australian marine ecosystems
- To develop the best possible technological base, analytical tools, and information resources to aid in the conservation, development, and management of tropical marine ecosystems

The Tropical Oceanography Program seeks to improve our understanding of the physical, biological, and chemical processes which operate on and in the coastal, shelf, and oceanic waters bordering northern Australia. Oceanography is a multidisciplinary science. Finding solutions to many oceanographic problems requires an inter-disciplinary approach. Oceanographic phenomena and processes are variable in time and hierarchical in structure, with local or regional phenomena being forced by both periodic and episodic processes operating over large spatial and longer temporal scales. The spatial distribution and magnitude of virtually all biological and chemical processes in marine ecosystems are coupled in some fashion to water circulation. Conversely, selected biological and chemical species, compounds or processes can sometimes be used to deduce the nature or magnitude of physical processes not accessible to direct observation.

A primary goal within the program is to quantify and model water circulation at local, regional, and basin scales. Building upon research activities started in the previous quinquennium, considerable effort will be directed toward understanding the two- and three-dimensional dynamics of flow within the complex matrix of the Great Barrier Reef and along its coastal and oceanic boundaries. Local boundary layer phenomena are being studied from the perspective of wave mechanics and turbulent transport.

The productivity of coastal, shelf, and oceanic marine ecosystems is related to the source, fate and cycling of carbon and other nutrient elements. Research activities in biological and chemical oceanography will focus upon exchanges of carbon between the atmosphere, water column, pelagic biota, and benthos in tropical shelf ecosystems and the role of nutrient controls at the system scale. These efforts will build upon the results of long-term research activities started in the previous quinquennium and make a significant contribution to Australia's involvement in global carbon cycling studies.

The biology, dispersal, and stock recruitment of large pelagic fish in the western Pacific ocean, tropical invertebrate species, and the pelagic larval stages of reef fish are still poorly known. Important life-history processes are directly affected by physical mixing and dispersal processes at local, regional, and shelf scales. Building upon research initiated in the previous quinquennium, projects in fisheries and population genetics will put aspects of the biology and biological oceanography of selected commercially or recreationally important species onto a quantitative footing through ecological studies of population dynamics and genetic analyses of population structure. Particular emphasis will be given to the relationships between ecological and physical processes at appropriate scales.

Throughout the quinquennium, research toward the solution of particular problems will require inputs from several disciplines. To formalise these interactions, generalise the results and explore their implications in a quantitative fashion, members of the program will seek to develop appropriate numerical analyses and models of phenomena and processes.

The Tropical Oceanography Program manages three research support facilities: the Remote Sensing Facility, a network of remote weather stations on the Great Barrier Reef and in adjacent control habitats, and an oceanographic instrumentation maintenance facility.

Overview

The research emphasis of AIMS is clearly directed to the diverse ecosystems which characterise the continental shelf regions of the coastal countries of the tropics. The methodologies that we have adopted to study these complex ecosystems are equally applicable to temperate waters. The results of the research are most directly applicable to the needs of the developing countries of the tropical regions of the world and AIMS is committed to ongoing collaborations, to working with these countries, and to maximizing the prospects of ecologically sustainable development of marine resources. The increasing significance of the United Nations Conference on the Law of the Sea and the declaration of Exclusive Economic Zones by an increasing number of countries draws particular relevance to our studies. As we move to the mixture of fundamental research and the applications of the results of that research to the benefit of the Government of our countries and to other countries and to commercial and industrial exploitation for economic return, it is essential that the Institute progressively be more involved with managers, industrialists, economists, decision makers, and international legal experts. We do not see this as a challenge to the integrity of the scientific research of the Institute. We do see it as an opportunity to interact more closely with institutions throughout the world which are involved in the understanding of marine processes and in working towards a global policy of ecologically sustainable development of our marine resources.

Canada

The Bedford Institute of Oceanography: Current Program and Future Directions

STEPHEN B. MACPHEE[1]

Key words. Oceanography — Marine science — Hydrography — Bedford Institute of Oceanography — Marine biology — Coastal — Zooplankton — Aquaculture — Operational oceanography — Atlantic Geoscience Centre

Summary. The Bedford Institute of Oceanography (BIO) and its two associated laboratories, St. Andrews Biological Station and the Halifax Fisheries Research Laboratory, are funded by the Government of Canada through the Departments of Fisheries and Oceans, Energy, Mines and Resources, and Environment Canada. The Department of Fisheries and Oceans (DFO) is the proprietor of BIO and the largest participant in the research program. The overall objective of the science program at BIO is to ensure that scientific information of the highest standard is available to the Government of Canada for use in developing policies, regulations, and legislation regarding the oceans and the living and nonliving resources contained in the oceans and in the sedimentary rocks below. There is a role to play in making this information available to the ocean industry and to the general public.

BIO is unique as a scientific institution in combining all disciplines of marine science in one location. Scientific teams investigate the physical and chemical properties of the ocean, the life within it, and the geology of the sea floor, the interactions between the oceans and the atmosphere and the oceans and the continent, and develop new methods and technology for investigating present day processes and past history. Institute scientists collaborate extensively with university, private industry, and government colleagues, nationally and internationally, by publishing in the primary literature, producing atlases, charts and other publications, transferring technology, and providing scientific advice.

With respect to future directions, our scientific program will continue to be multidisciplinary and will embrace not just knowledge available within national boundaries but global knowledge on important scientific issues. The trend

---

[1] Department of Fisheries and Oceans, Bedford Institute of Oceanography, P.O. Box 1006, Dartmouth, Nova Scotia, Canada B2Y 4A2

towards major field experiments and numerical experimentation will continue through participation in projects such as the World Ocean Circulation Experiment (WOCE), the Joint Global Ocean Flux Study (JGOFS), and Fisheries in a Changing Climate (FICC). Some of the specific laboratory themes will include (1) more emphasis on the development of instrumentation for acoustic studies on zooplankton biomass and production, (2) development of oceanographic data products to aid in stock assessment and monitoring environmental change, (3) cold water aquaculture diversification, (4) remote sensing, (5) ocean modelling, (6) improved estimates of carbon fixation and vertical fluxes in the North Atlantic, and (7) shoreline erosion studies.

Evaluation of hydrocarbon and mineral resource potential in the offshore will increasingly require sophisticated modelling to obtain an understanding of resource generation and accumulation. Research aimed at an improved understanding of the marine environment will allow society to make wise decisions on the exploitation of these resources and the preparation of appropriate scenarios for land use management and development of Canada's offshore lands.

Introduction

Economic Perspective

Canada (Fig. 1) is a coastal nation bordered by three oceans. It has the longest coastline (244,000 km) of any nation and a continental shelf of area more than one-half of the Canadian land mass. The area inside Canada's 200-mile zone is, in fact, $3.26 \times 10^6 \, km^2$ or 30% of the Canadian land mass. In addition, with less than 0.5% of the world population, Canada has 16% of the world's total surface area of fresh water.

The ocean economy, from the fisheries, marine transportation industry, offshore manufacturing and service sectors, annually contributes C$6.5 billion to the country's gross domestic product (GDP). Another C$4 billion is contributed in spinoffs from the recreational fishing industry. Approximately 160,000 direct jobs are created by these ocean industries. Over the next decade, it is hoped that increased hydrocarbon extraction and aquaculture will add C$2–3 billion to this total.

The present value of the nation's capture fishery exceeds C$3.2 billion/annum (1990 figure) and this industry provides employment to about 86,000 fishermen and 38,000 plant workers in regions of Canada where few opportunities for alternative employment exist. The 1990 value of the aquaculture fishery was C$167 million. The potential for aquaculture in Canada (both finfish and molluscan) is very much underutilized and this industry is expected to reach from C$500 million to C$1 billion/annum by the year 2000 [1].

Marine transportation is also an important industry. Canada is a major trading nation and derives 25% of its GDP from external trade, of which 55% is carried by sea. Most of the major cities are coastal and many are also major

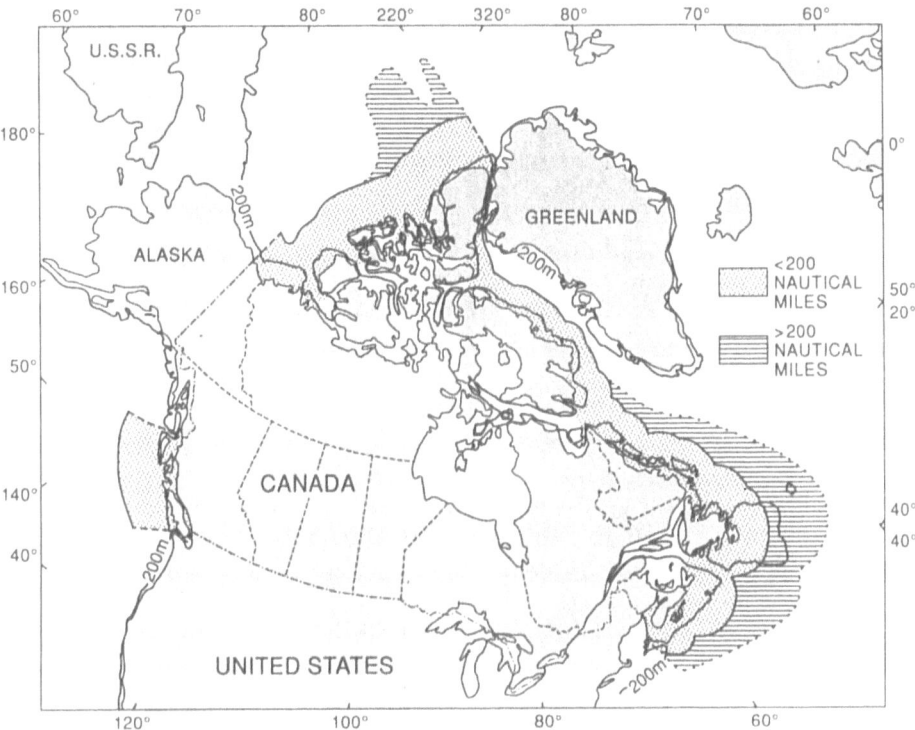

Fig. 1. Map of Canada including continental shelf

seaports, with 15 of these seaports having exports of more than 1 million tonnes of cargo/annum. Offshore exploitation has been a cyclical industry but it is currently showing positive signs with two production facilities under construction. This industry is expected to inject C$1 billion/annum into the economy by the year 2000. Interest in coastal recreational activities, such as angling, boating, sailing, whale watching and diving, continue to grow quickly in Canada.

There are some of the commercial reasons why the oceans are important to Canada and to Canadians. In the remainder of this paper, I will attempt to explain the scientific rationale for Canada's marine science program, as well as some of the major thrusts.

Scientific Perspective

To date, Canada has focused the major component of its marine science and marine-science-related activities in government laboratories. This trend is slowly changing, however, with a focus on partnerships (university, government, and the private sector), an emphasis on having more of the basic research carried out in universities, and considerable encouragement being provided to increasing the level of private sector research.

A number of government departments carry out marine research, with the Department of Fisheries and Oceans (DFO) being the major player in the disciplines of biological sciences, physical and chemical oceanography, and hydrography. The Bedford Institute of Oceanography (BIO) is the largest DFO laboratory. The other main federal oceanographic research laboratories are: the Pacific Biological Station (PBS), Nanaimo, British Columbia; the Institute of Ocean Sciences (IOS) at Sidney, British Columbia; the Maurice Lamontagne Institute (MLI) at Mont-Joli on the Gulf of St. Lawrence in Québec; and the Northwest Atlantic Fisheries Centre (NAFC) in St. John's, Newfoundland.

The Department of Energy, Mines and Resources (DEMR) is responsible for the Geological Survey of Canada which, as part of its mandate for understanding the geology of Canada, carries out marine geoscience research and surveys in Canada's coastal and offshore regions. The Department of the Environment (DOE) carries out air-sea interaction research as part of its marine weather, marine and ice forecasts, and climate prediction. DOE also carries out mission-oriented research for the protection, conservation, and preservation of the marine environment. DOE has two small laboratories at BIO. Several other government departments, including the National Research Council, are involved in ocean science research, but to a lesser extent.

University scientists obtain some funding for their research programs from the universities and the private sector, but the major funding is through government-managed subvention programs. The emerging trend is for government to support major science programs where two or more universities are networked with the private sector and with government research establishments to carry out large research projects. The main universities involved in marine science in Canada are: Memorial University, St. John's, Newfoundland; Dalhousie University, Halifax, Nova Scotia; Université du Québec à Rimouski, Rimouski, Québec; Université Laval, Sainte-Foy, Québec; McGill University, Montreal, Québec; University of British Columbia, Vancouver, B.C.; and University of Victoria, Victoria, B.C. Most at-sea components of university science are carried out from vessels owned and operated by DFO.

The high technology, private sector ocean industry has been in place for the past 20 years or more and has produced a number of notable successes. Government laboratories have been able to assist this industrial development by the transfer to the private sector of technology that has been developed to meet specialized needs. The private sector then provides the necessary re-enginerring for production to meet client requirements. The ocean industry remains a capricious commercial undertaking because many of the methodologies are still at a research stage, the clients are demanding, and the market is small. Total ocean manufacturing and service sector employment in 1989 amounted to 32,000 person years, with sales of C$1.8 billion [2].

Science Program-BIO

General

In this section of the paper, I will describe the organization and programs of BIO and two affiliated laboratories: the St. Andrews Biological Station, St. Andrews, New Brunswick (staff of 95), and the Halifax Fisheries Research Laboratory (staff of 55). These two laboratories are under the same management as BIO and carry out research that is fully integrated, so in the remainder of this paper when discussing the science, it can be assumed that I am including the programs of all three laboratories.

BIO, located in Dartmouth, Nova Scotia, is a Government of Canada establishment. The DFO, with more than 550 scientists, engineers, hydrographers, and support staff in the laboratories and on the ships, is the largest component of the research program. The DEMR component, with a complement of 120, is the largest marine geoscience laboratory in Canada. The DOE component (staff of 24) consists of a small analytical testing laboratory and a seabird research unit. In the following paragraphs, I will provide additional information on the various research disciplines and discuss the networking between the Institute scientists and the private sector and university communities.

Department of Fisheries and Oceans Program

Biological Sciences

The Biological Sciences program is a multifaceted program with the following principal thrusts:

1. *Biological oceanographic research* is carried out on the dynamics of marine eco-systems in coastal, shelf, and deep-ocean waters, with special emphasis on the interdependence of biological communities, their temporal and spatial variability, and their relationship with the physical and chemical conditions of the marine environment. Some of the more important projects currently underway include: (a) studying the biological influences on light transmissions in the ocean; (b) utilizing remotely sensed data with a view to obtaining information on phytoplankton biomass and production; (c) investigating the role of pelagic biota in the global CO_2 cycle; and (d) providing Canadian input to international scientific programs such as the Joint Global Ocean Flux Study (JGOFS) and the World Ocean Circulation Experiment (WOCE).
2. *Habitat ecology research* is carried out to ensure that healthy fish habitat is maintained, that damaged habitat is restored, and that new fish habitat is created where the production of fisheries resources can be increased. Major long-term research themes include: benthic ecology, benthic-pelagic exchanges, research on harmful marine phycotoxins, impacts of aquaculture, ecosystem analysis, atmospherically transmitted contaminants, and sublethal effects of offshore hydrocarbon development.

3. *Fisheries research* includes study of the biology and production processes of groundfish, pelagic fish, invertebrates, and marine mammals in Atlantic waters. This includes defining the population status of important commercial species, and determining the environmental, biological, and human factors causing the observed changes in these populations. The required data are obtained from research vessel surveys to sample fish stocks, as well as commercial fisheries data to provide independent indices of stock abundance. One of the main products of this research is the provision of biological advice on the total allowable catches for the various species.

4. Aquaculture research consists of studies on both finfish and shellfish. Finfish studies have centered around salmon physiology and genetics, fish diseases, and nutrition, mainly in support of sea cage culturing. Shellfish aquaculture research has included studies on the effect of harmful marine toxins, the accumulation of such toxins in cultured organisms, disease control, diets, and husbandry. Research is also being carried out on the effects of pollutants on aquaculture operations, the effects of waste accumulation on water supply, the flushing rates in areas intended for or in use for aquaculture, and the environmental impact of feed and drug residues.

Physical and Chemical Sciences

In the Physical and Chemical Sciences Program, research efforts are devoted to the following primary areas of research:

1. *Ocean climate*. Oceanographic studies are carried out on those processes that govern ocean and shelf circulations and the distribution of physical and chemical properties that are considered important to long-term needs for oceanographic data. This research includes large-scale modelling of the coupled ocean/atmospheric system and regional models important to fisheries, transportation, and pollution issues, and includes the Canadian contribution to WOCE.

2. *Marine developments*. Oceanographic studies are carried out to provide guidance to the safe, economic and legal management of engineering developments in the offshore. This includes the protection of human health, vessels, structures and equipment, and the marine environment.

3. *Living resources*. Oceanographic studies are carried out in collaboration with biological and fisheries scientists to further the understanding of the relationship between the environment and living resources in the ocean.

4. *Biogeochemistry*. Oceanographic studies are carried out on the processes that govern the distribution, fluxes, and properties of chemical parameters. This includes the behavior of both naturally occurring and anthropogenic compounds in the marine environment that may be transferred across the air/sea or sea/sediment interface.

5. *Toxicology contaminants and habitats*. Research is carried out to provide advice on environmental or fisheries issues arising from concerns over chemical and other marine contaminants. This includes the effects on fish production and quality and upon the marine environment and its users.

Hydrography

The Hydrography Branch is charged with carrying out surveys to provide navigation charts for commercial navigation and fishing and for recreational boating. In addition to navigation charts and other publications such as Tide and Current Tables, Sailing Directions, and Small Craft Guides and similar publications for the mariner, Territorial Sea and Fishing Zone Charts are produced under the hydrographic program. Under a cooperative program with DEMR, data are gathered to produce offshore maps depicting the gravitational and magnetic fields as an offshore extension of the terrestrial mapping program. The Branch also carries out an extensive research and development program aimed at accelerating the survey and chart production program and transferring relevant technology to Canadian industry.

Department of Energy, Mines and Resources Program

The second largest science grouping in the Institute is in the Atlantic Geoscience Center, part of the Geological Survey of Canada, one of the sectors of the DEMR. This laboratory provides knowledge, technology, and expertise concerning the non-renewable resources of the Atlantic and Arctic offshore regions of Canada.

One of the major goals of this laboratory is to delineate the crustal and lithospheric framework of Canada's continental margins in an attempt to understand the processes which led to the formation of these modern margins and their sedimentary basins. In addition, research is being carried out to improve the understanding of the internal geology of these sedimentary basins and their hydrocarbon resources.

A second major goal of this laboratory is the study of the quaternary geology and the engineering geology in coastal and offshore regions. This research is carried out to provide a coherent picture of the nature and distribution of sediments on the eastern Canadian and Arctic margins of Canada and their contiguous ocean basins. Included, as well, is the study of the physical properties of seafloor sediments and the modern processes which modify these properties, as required for the solution of engineering and land use problems. The research program is very much aligned with the development of new technology and adaptation of existing technology to enhance capabilities and open new opportunities.

All these activities are carried out in close collaboration with other branches of the Geological Survey of Canada and with scientists in DFO, the universities, and the private sector. In fact, many research projects are multidisciplinary in scope involving government, university, and private sector scientists and engineers working together in the laboratories and on the research vessels.

Department of the Environment Program

The DOE, the third government department to have staff housed at BIO, has one laboratory consisting of approximately 14 scientists and support staff involved in carrying out chemical, toxicological, and microbiological analyses in support of the regional Environmental Protection Services Program. A Seabird Research Laboratory of approximately ten scientists and support staff is also located at the Institute.

Networking with the Private Sector and Universities

Institute scientists and engineers design instrumentation and software, not readily available commercially, to carry out research programs and to interpret large data sets. If the instrumentation or software is commercially viable, it is turned over to the private sector to be engineered to a production standard and is then available to be marketed commercially. To assist in this technology transfer, some space is leased to high technology commercial companies which then have the opportunity to work closely with Institute scientists. BIO is not a degree-granting institute but scientists are provided the opportunity to supervise graduate students in local universities and to hold adjunct professorships. In addition, many Institute scientists collaborate with university scientists and share shiptime with them on research cruises.

Future Directions in Marine Science

General

It was just a little more than 10 years ago that work began on the report "Ocean Science for the Year 2000" [3]. From a recent review of this report, it is quite apparent that the questions posed in each subject area in 1980 are still applicable, i.e., (1) what are the most important ocean research questions that should receive particular attention in the next decade, (2) what are the major advances to be expected and what kind of research should be encouraged for them to be achieved, and (3) what are the principal impediments to achieving these advances. Indeed, while considerable advances have been made in all the marine science disciplines during the past decade, many of the particular programs underway in 1981 are still ongoing. Probably the largest changes that have occurred in the past 10 years have been in the area of instrumentation and in the use of larger computers for numerical modelling. Other areas of rapid change include an increased focus on global experiments, an increase in interdisciplinary research, and greater awareness of the societal impacts of ocean research.

There will be an equal or greater number of challenges and responsibilities in the marine sciences in the next 10 years. One of the challenges currently being faced in a number of countries is the network of private economic rights that is extending into the offshore. These rights include the granting of individual rights to fishermen and to fishing companies, the granting of oil and gas leases, aquaculture permits, ocean dumping permits, and the establishment of exclusive tanker zones. The demands of these stakeholders will in future cause disagreement and a difficult resolution process in both a national and international sense.

Other challenges that come to mind include (1) conserving ocean resources and the marine habitat, (2) managing conflict among the various stakeholders in the coastal zone, (3) understanding and managing the impacts of climate change, and (4) dealing with fish diseases for cultured stocks and the effects of harmful marine toxins. There will also be many opportunities to carry out research more efficiently as satellite data and supercomputers become more readily available and as new instrumentation is developed and applied. In this paper, rather than discuss "Future Directions" in a general sense, I have decided to select five specific research projects and discuss these projects in some detail. The five projects I have selected are: (1) From Oceanographic Research to Operational Oceanography, (2) Cold Water Aquaculture Diversification, (3) Measuring Shelf Zooplankton Acoustically, (4) Acquisition and Management of Larger Bathymetric Data Sets, and (5) Environmental Responsibility in the Coastal Zone.

From Oceanographic Research to Operational Oceanography

One of the challenges faced by a research laboratory is the provision of information and data products to users outside of the immediate scientific team, to other organizations and to other disciplines. As the amount of research data increases, there is a growing demand from client agencies for useful products that will assist them in solving problems such as assessing the abundance of fish stocks, determining suitable sites for aquaculture facilities, or modelling the effect of an oil spill. Aircraft and satellite remote sensing and modern instrumentation have rapidly increased our ability to acquire data and modern computers provide a dramatic increase in the speed in which data can be analyzed. The availability of high speed data networks allows easier access to data but also introduces a new level of complexity in that the data and software needed for a particular project are frequently spread over a number of different machines. At BIO, a major thrust is in the development of the necessary hardware and software infrastructure to facilitate third party access to data, to enhance analysis expertise, and to develop new oceanographic products for the client community.

Providing access to data is a multifaceted problem. To inform users of data availability and how these data may be accessed, a series of on-line directories are being developed describing all the physical oceanography holdings in the laboratory. Since the client community is equally divided between those requiring immediate statistical summaries and those requiring access to complete data sets to carry out their own analyses, the directories will contain detailed monthly summaries of means and variances as well as the ability to download data sets to a format suitable to the client requirement. Upon completion of these directories, the next step will be to expand them to include modelled parameters which are becoming a useful data source as the sophistication and reliability of numerical models improve.

Combining data from different data sources in order to subject them to a uniform analysis is often made more difficult by having to deal with a variety of formats and data structures. To facilitate this process, there has been a major effort over the past 2 years within the physical and chemical science disciplines at BIO to develop an analysis and data management system which is a radical departure from previous systems. Instead of developing large programs which perform specific tasks and require a rigid input and output format, the analysis software, written to strict ANSI standards, consists of a number of modules or functions which operate independently of any data format. The modules can be assembled in various ways to provide the desired analysis. As new analysis techniques are developed, modules can readily be added to the system without affecting the ones already existing. The data management component, which is seen by the system as simply another module, permits scientists to define the structure of incoming data to the analysis routines. Using this approach, data from a variety of sources can be combined and subjected to the same uniform analysis without the scientist having to write any new software. The system, which to date consists of over 125 oceanographic modules, has been received with a great deal of enthusiasm within Canada and interest has been expressed from research laboratories in the United States, Korea, and China.

Incorporating observational data into statistical or numerical models of the same fields is another way in which attempts are being made to maximize the information available from existing data, as well as to determine how much additional data are required for a given level of accuracy. The process of combining observational data with models, called data assimilation, has been in use for some time in the study of meteorology, but in oceanography the methodology is still being developed. The imminent availability of oceanographic satellite measurements that will cover the globe in both space and time, however, is accelerating the efforts at BIO to make use of these techniques. Various aspects of the research include developing algorithms to produce surface wind and pressure fields from satellite imagery, combining these data with *in-situ* observations, and coupling both types of data into models. Optimal estimation techniques are being investigated to generate gridded data fields from observational data which are generally non-uniform in space and time and have different resolutions and accuracies. Sophisticated data sorting algorithms are being developed to allow efficient access to large data sets, a necessary step before considering the large volumes of modelled parameters as a source of data to be managed and archived.

With an understanding of a system that comes from the analysis of both observational data and the use of modelling techniques, it becomes possible to define oceanographic, hydrological, and meteorological indices which may then be applied to a whole range of problems from long-term climate change to shorter term fluctuations in abundance of a particular fish stock. By comparing these oceanographic indices with individual sites, monitoring locations can be established to reduce the need for massive observation programs.

As an example of the techniques discussed, a scientific study is currently underway to determine the dominant temporal and spatial scales of variability in the Scotian Shelf region. The intent is to identify the primary forcing functions

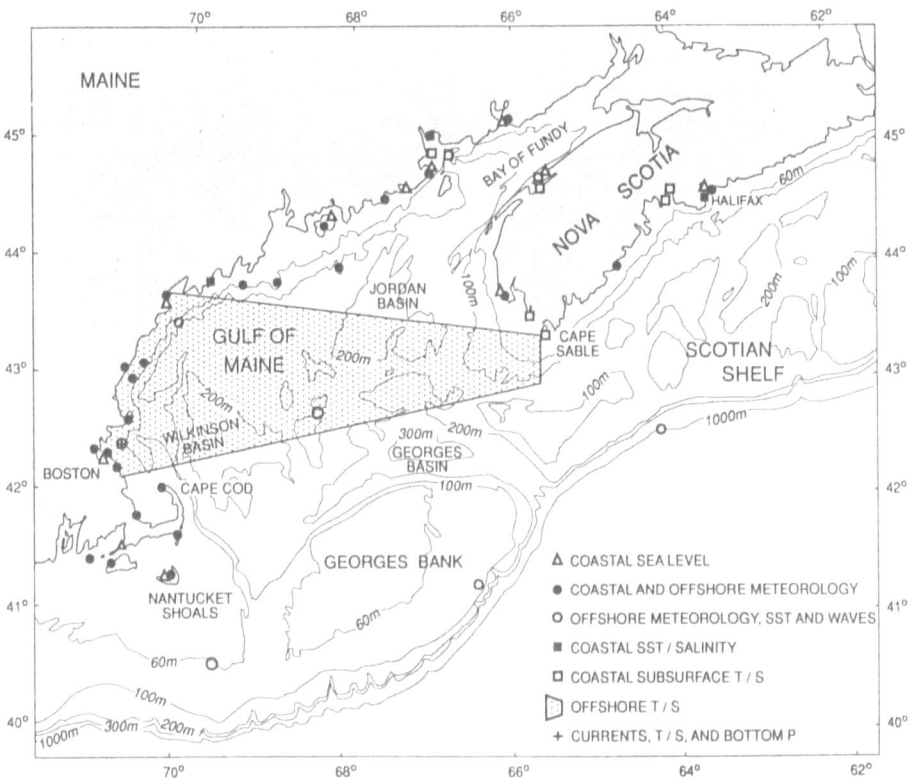

Fig. 2. Map of Gulf of Maine region showing the location of present monitoring sites for Physical Variables

controlling climate change and to determine what large-scale environmental factors may be at least partially responsible for long-term fluctuations in ground-fish abundance. The Gulf of Maine area (Fig. 2) has been selected because of its relatively high variability and its importance to the fisheries. In total, some 55,000 stations with over 600,000 observations of temperature and salinity were obtained from both the Canadian Marine Environmental Data Service (MEDS) and the U.S. National Oceanographic Data Centre (NODC). When the database is expanded to include the entire shelf region there will be approximately 500,000 stations and 5 million observations. Twenty-five long time series of various oceanographic and meteorological parameters have been assembled into the database. Some of the series, such as sea surface temperature off Halifax Harbour and St. Andrews, N.B., span 7 decades [4].

When the system is fully operational, BIO scientists will initially select and review data using a high speed data link to query the MEDS databases directly. As with any large database, identifying duplicate or erroneous data is a major problem. Quality control tests developed by MEDS in conjunction with the

Global Temperature-Salinity Pilot Project (GTSPP) have been adopted to further quality control the data.

A simple data structure consisting of the space-time coordinates (latitude, longitude, depth, time), parameter name, and parameter value has been defined. This structure permits us to readily move the data between the physics analysis system for computation of the climate indices and a relational database (ORACLE) for incorporation with biological sciences stock assessment programs. Selected volumes of data are transferred over the network to a graphics mini-supercomputer and grid interpolation performed using nearest neighbour and optimal estimator techniques. The same machine will be used to produce three-dimensional color displays of the data and animated movies. When this project is complete in about 4 years' time we will have developed all of the components of a system that will integrate data acquisition, quality control, volume selection, statistical analyses, and 3D graphical displays of the ocean.

The analysis and interpretation techniques which are now being investigated and developed represent one of the most exciting developments taking place in oceanographic research in our laboratory. It brings together scientists from many disciplines and will lead to an understanding of oceanographic processes on spatial and temporal scales not previously possible.

Cold Water Aquaculture Diversification

As mentioned earlier in the paper, aquaculture is a fast growing industry in Canada with Atlantic Canada being one of the areas of recent growth and an area with considerable potential for increased production. Most of this production will occur in estuaries, inlets, and coastal embayments. While both finfish and shellfish farming are rapidly developing in many areas, the principal salmon aquaculture area is located in the lower Bay of Fundy along the south-western coast of New Brunswick (Fig. 3).

Within this region the most intensively used area is the Letang Inlet system, primarily because of two environmental conditions: (1) the relatively intense tidal mixing of the water column is sufficient to maintain winter minimum temperatures above the lethal level ($-0.7°$C) for Atlantic salmon (*Salmo salar*), enabling them to be held in cages over an 18-month growout period; and (2) the presence of several islands provides protection from open-ocean waves and hence prevents damage or destruction of the cages.

In cage culturing of finfish, food is supplied and, therefore, is not a limiting factor. Production is limited by a "holding capacity," i.e., the potential maximum production is limited by a non-trophic resource (e.g., dissolved oxygen availability). The ability of the environment to accept waste products from an aquaculture site is dependent on its "assimilative capacity," i.e., the degree to which wastes can be absorbed and dealt with locally.

The holding and assimilative capacities of a given inlet or embayment are determined by a wide range of physical, chemical, biological, and sedimentological processes which must be considered in a holistic fashion. For example, the

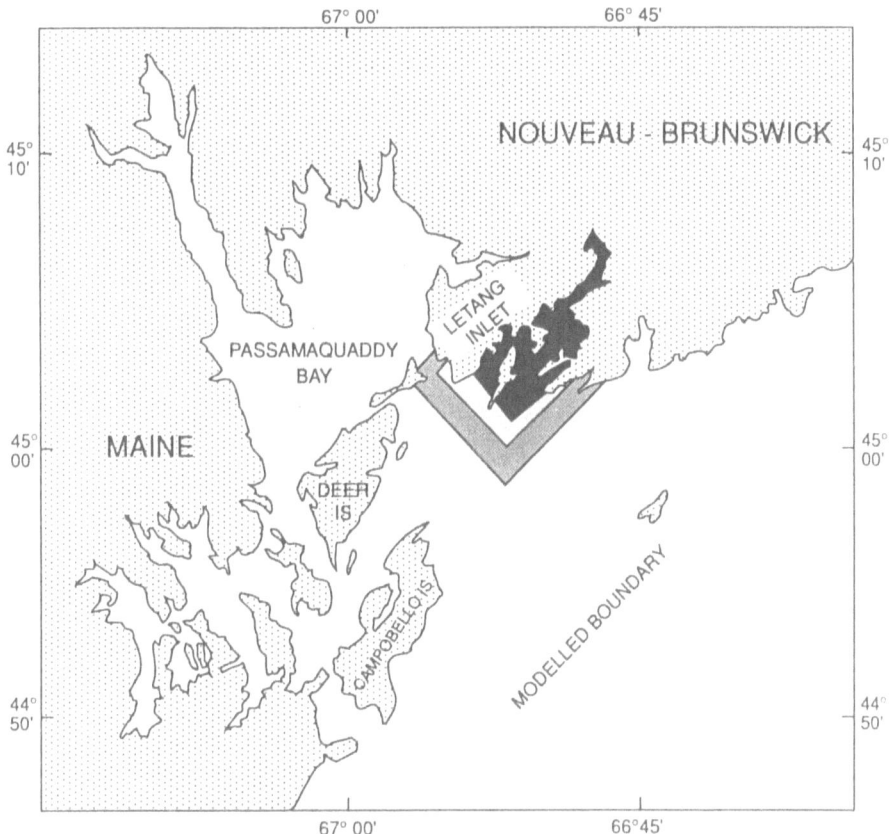

Fig. 3. Map of southwest coast of New Brunswick showing Letang Inlet

transport of dissolved and particulate aquaculture wastes away from a cage site is controlled by currents and sedimentation processes, their composition is altered by microbial and chemical transformations, and their ultimate impact is often expressed in terms of effects on commercially important wild fisheries such as lobster or herring. A multidisciplinary approach to studying the environmental impacts of aquaculture is therefore essential.

Growout of salmonids in the sea is an intensive form of monoculture with biomass reaching $15-20\,kg$ wet fish m^{-3} of seawater within the cages. High densities of salmon, growing at the maximum rate, produce large quantities of wastes. The wastes include dissolved and particulate components consisting of feces, ammonia, and uneaten food. The chemical composition and amounts of these substances depend on the type and amount of food supplied to the fish. Both the food and excretory wastes contain nitrogen and phosphorus compounds which are plant nutrients in the dissolved state and thus may stimulate eutrophication. Dissolved oxygen is also removed from the water both by the

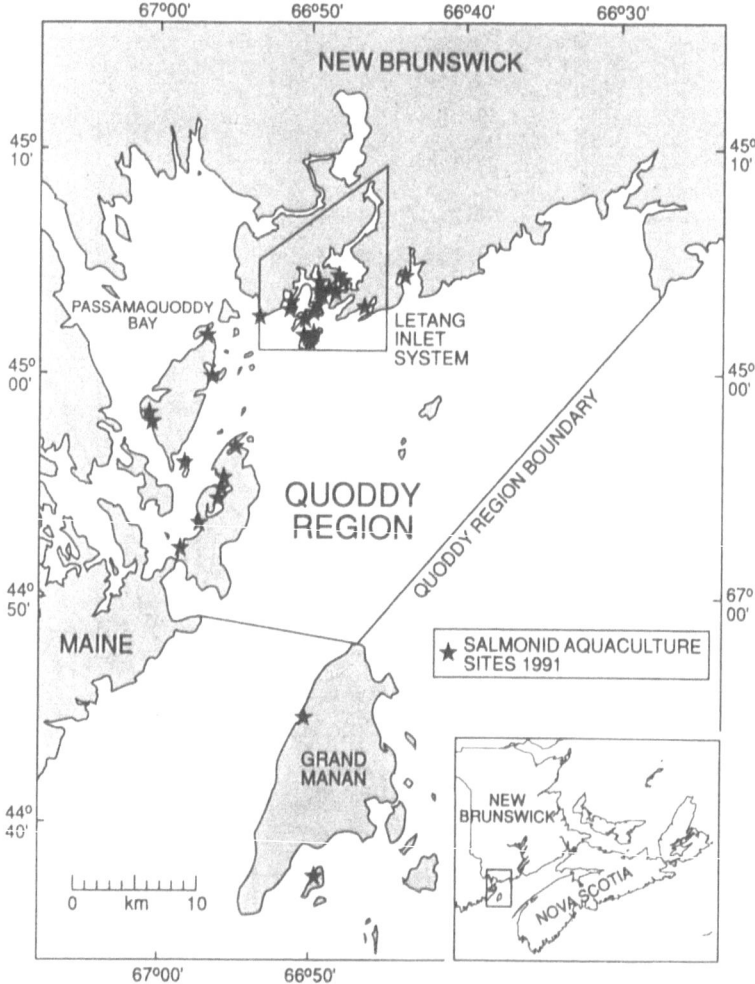

Fig. 4. Map of Letang Inlet system showing salmon aquaculture sites

salmon and by the decomposing wastes (biological oxygen demand or BOD) whether in the water or on the bottom.

Particulate food and fecal wastes usually result in increased levels of sedimentation beneath or near the cages. The environmental factors which influence how much of this material reaches the sediments directly under the cages depend on the strength of current, the rate of supply of new water, the depth of the site, the composition, size and behaviour of the particulate matter released, the temperature and salinity of the water, as well as the timing and occurrence of wave actions which result in resuspension and transport of previously settled sediments. Where a buildup of wastes occurs, a loss of aerobic microflora and most macrofauna occurs because of the lack of sufficient dissolved oxygen or presence

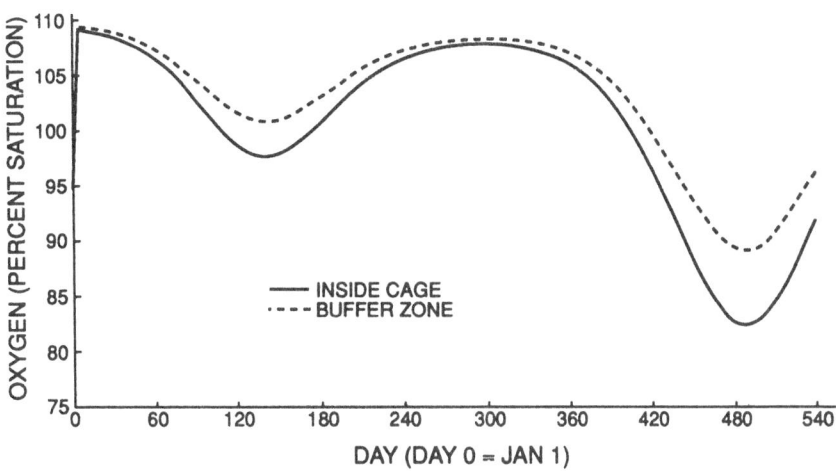

Fig. 5. SITE model showing simulated levels of dissoled oxygen inside fish cages as well as in the surrounding $100 \, m^2$ buffer zone, during 18-month growout period

of toxic substances, and hydrogen sulphide or methane will be released from the bottom. The presence of a polychaete-dominated, low diversity, benthic community is a strong indication of an organically polluted benthic environment. If the buildup continues the polychaetes also disappear and sulphur oxidizing bacteria, usually *Beggia-toa* sp., will be the only life forms present. These bacteria can exist only at the boundary layer between oxic water and anoxic sediments below, and will develop into large mats immediately below a cage site.

The multidisciplinary team approach to study the Letang Inlet system and the environmental impacts arising from the rapidly developing salmonid cage culturing, involves both field measurements and modelling exercises. A two-dimensional tidally driven numerical model has been developed to provide detailed information on current problems and flushing characteristics of the Inlet system. The nested grid cell sizes vary in scale from 800 m for Passamaquoddy Bay and the offshore area down to 100 m for the Letang Inlet system where detailed information is needed (Fig. 4). Model simulations can be undertaken using input of wastes (BODs, nutrients) from existing salmon farms and the resulting distribution of enhanced nutrients and dissolved oxygen deficits mapped for any time of the year. Additionally, other local wastes sources (e.g., fish processing plants, pulp mills) can be included as well as the effect of any proposed additional salmon farms examined.

Concurrently, a multidisciplinary team has developed a model called SITE [5] which simulates the most important biological processes that take place within a fish cage as well as the physical flushing processes. It calculates the consumption of oxygen (Fig. 5) and release of wastes on approximately an hourly basis during the 18-month growout period and estimates the concentrations both in the cage and surrounding waters. The output of this model is used to determine the loadings for the detailed water quality model.

Closely integrated with these numerical modelling studies have been a large number of field investigation in the Letang Inlet system which have collected data on currents, diffusion, flushing, phytoplankton abundance and species composition, water chemistry, suspended particulates, surficial sediment composition, sediment chemistry, and benthic/pelagic fluxes. These field programs have been directed in part by the data requirements of the modelling exercises and the results used to further develop and refine models which ultimately can be used jointly by scientists, environmental managers, and growers to estimate and predict the likely environmental impacts of both existing and proposed aquaculture facilities in coastal habitats. Through this approach, it is hoped that the aquaculture industry can continue to develop in an environmentally acceptable manner but yet stay within the limits imposed by natural processes. In the coming years, the expertise developed for the Letang Inlet will be transferred to other Canadian coastal areas experiencing similar development pressures.

Measuring Shelf Zooplankton Acoustically

Since the first patent was granted for echo sounding in 1907, acoustic techniques have become increasingly more important in oceanographic and fisheries research. While this importance will continue to accelerate in areas ranging from bathymetric surveying to assessing the biomass and migration of fish stocks, to measuring ocean currents, to geophysical surveying and even to climate prediction, the single acoustic application we wish to focus on in this paper is the use of acoustics in the measurement of zooplankton.

Zooplankton distribution on our continental shelves are generally controlled by the prevailing currents which generate their seasonal cycle. In addition, many continental shelves contain deep basins which harbour nearly all the zooplankton biomass during winter. The problem we are attempting to resolve is to determine how zooplankton are distributed throughout the Canadian Atlantic shelf, their species composition, the physical processes governing their distribution and their role as prey to fish populations. In addition, it is important to understand how climate change will affect these ecosystems and how to monitor the long-term changes in abundance and species composition.

Acoustic techniques offer a number of advantages over other methods in achieving this goal as they are non-intrusive, provide distributional data in real time, and provide high spatial resolution horizontally and vertically. In addition, acoustics can give researchers accurate information on the size and abundance of zooplankton (0.2–10 mm), microneckton (1–10 cm), and fish simultaneously. Within the last decade there have been important advances in the field of bioacoustics with the development of multifrequency and dual-beam systems for extracting zooplankton information from volume-backscat-tering data. Using multiple frequencies, acoustic size distributions can be extracted from volume

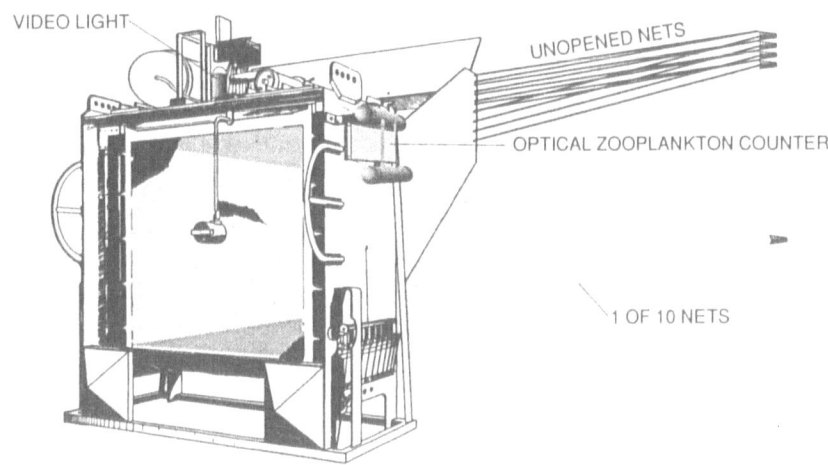

VIDEO LIGHT

UNOPENED NETS

OPTICAL ZOOPLANKTON COUNTER

1 OF 10 NETS

Fig. 6. BIONESS with one of its ten nets open

backscattering data collected at several different frequencies [6]. The intensity of the backscattering is a function of the number and the size of the organisms in that volume. Mathematical inversion models are used to relate the backscattering intensities from the different sizes of organisms for each frequency and thereby determine the biomass and quantity for various length classes of organisms.

The acoustic sampling system presently in use is a multifrequency system capable of operation at 12, 50, 120 and 200 Khz [7]. Groundtruthing and supplementary data are provided by the BIONESS [8], a multiple net zooplankton sampler with ten opening and closing 1-m nets controlled from the ship (Fig. 6). This sampler, towed at 2 m/s, includes a conducitivity, temperature, depth measuring instrument (CTD), a flow measuring device, an optical plankton counter, and a video camera and lights. It is used to collect organisms in a size range of 0.25–100 mm.

An additional sampling platform, the BATFISH, containing an optical plankton counter, a fluorometer, and a CTD (Fig. 7), is towed at speeds of 4–5 m/s along an undulating path from the surface to within several metres of the bottom to provide synoptic coverage of zooplankton distribution and layering over large shelf scales (Fig. 8). An *in-situ* optical plankton counter [9] provides size and concentration data for zooplankton in the size range 0.2–30 mm (Fig. 9).

An advanced design eight-frequency acoustic system operating between 50 Khz and 1 Mhz is currently being constructed. This instrument will be capable of operating at depths from the surface to 300 m with a sensitivity up to 100 times greater than the present multifrequency acoustic system. This new system will allow the detection of animals in the size range from large copepods (2 mm) to fish at orders of magnitude lower than any concentrations in the past. We will also be able to acoustically see and count the individual krill and thereby be able to more accurately size the animals.

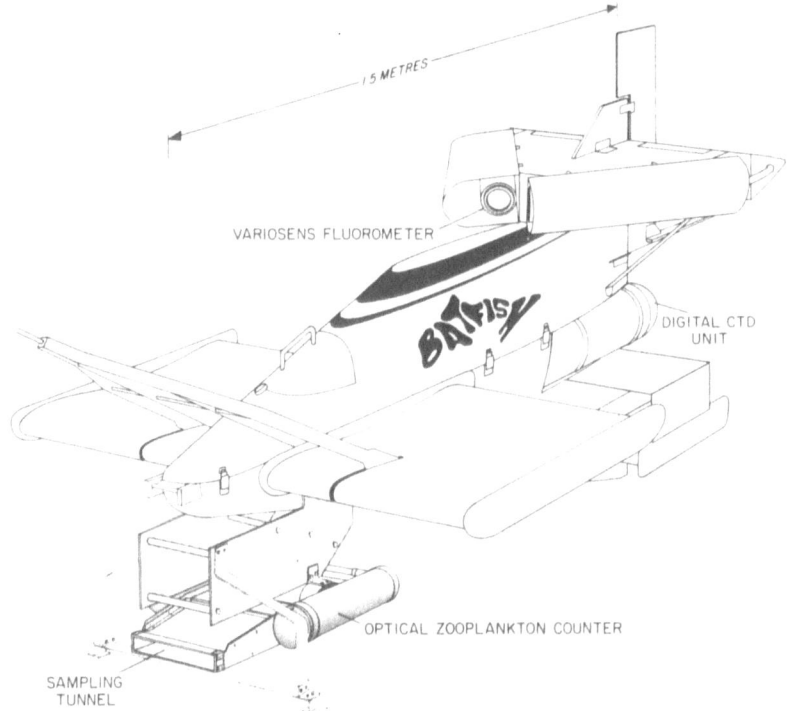

Fig. 7. BATFISH with sensors

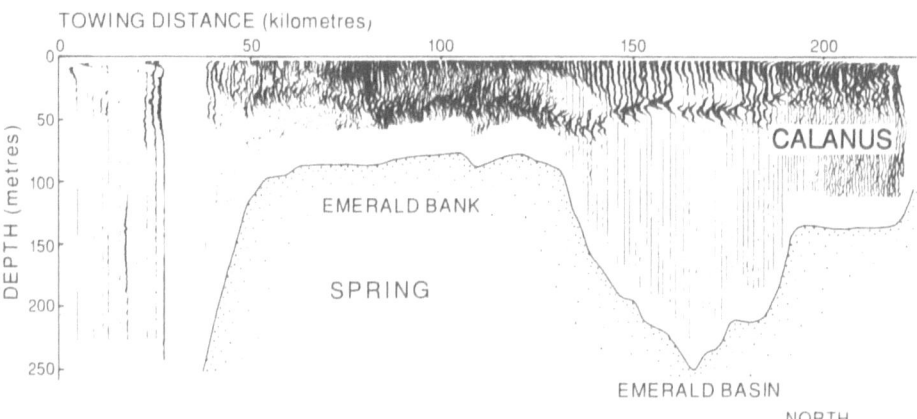

Fig. 8. Synoptic coverage of zooplankton distribution measured with optical plankton counter

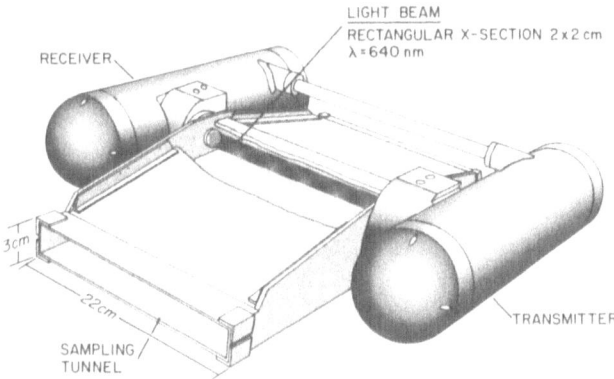

Fig. 9. Optical plankton counter

One of the greatest deficiencies in the present sampling system has been our inability to collect long-term time series data on the concentrations and distributions of zooplankton and microneckton. It is planned, therefore, to construct two new instruments for the collection of time series data for periods of up to 1 year. The first of these instruments will be a moored optical plankton counter that will collect data on the concentrations and sizes of zooplankton at a fixed depth near the bottom. The mooring (Fig. 10) will also include current meters. The second new instrument will be a botton-moored acoustical system will measuring frequencies of 200 Khz and 1 Mhz tuned to measuring euphausiids and copepods respectively. These two new instruments will provide longer time series data on the major biological changes in the zooplankton community and on the changes in the populations of krill and fish at different seasons and during physical events such as major storms or changes in water masses.

High resolution video cameras will be utilized extensively in future sampling of zooplankton. The cameras will be mounted on the BIONESS and BATFISH with video lights to observe the organism in front of the samplers to study their reactions to the sampling vehicles as well as provide data on their *in-situ* concentrations and distribution patterns. The video data will compliment the zooplankton samples collected in the nets and help in the identification and concentration estimates of animals responsible for the acoustic backscattering.

While this application represents but one use of acoustics in oceanographic research, nevertheless, it demonstrates the versatility of acoustic measurements in obtaining research data on marine ecosystems.

Acquisition and Management of Larger Bathymetric Data Sets

The practice of hydrographic surveying, mapping the sea floor or the floors of inland water masses by concurrently measuring depth and position, has changed dramatically over the years and particularly in the past 2 decades. While the

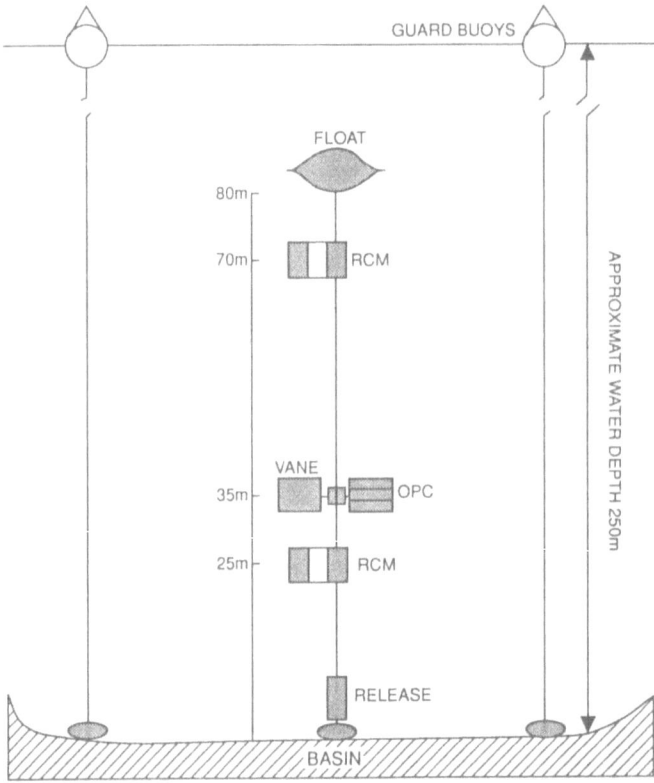

Fig. 10. Moored optical plankton counter configuration. *RCM*, recording current meter; *OPC*, optical plankton counter

methodologies have changed to provide data more efficiently and more rapidly, the demands generated by commercial, fisheries, and recreational vessels, marine resource exploration and exploitation, boundary delimitation, and defence requirements have also changed at an even more rapid pace. To satisfy these increasing demands in an efficient manner, hydrographers must continue to develop ways of gathering and processing data more rapidly and more accurately.

Up until the advent of the acoustic echo sounder, surveys were carried out by using a lead line to measure depth and visual or astronomic methods for the determination of position. This method was time consuming and shoal areas could easily miss being discovered. In deep water, the determination of depth was an arduous task taking a whole day to obtain each measurement. On the CHALLENGER expedition of 1872–76, stout hemp rope was flaked out on deck and allowed to run free over the ship's side. After it was determined that the weighted end had reached the sea floor, it was hauled back on board by hand over a steam capstan. It is interesting to note that by 1911, only 5,969 ocean soundings from more than 1,000 fathoms depth had been made [10].

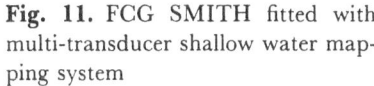

Fig. 11. FCG SMITH fitted with multi-transducer shallow water mapping system

While early echo sounding systems recorded a single line of depths directly beneath the ship's keel, with modern multi-beam and multi-transducer systems, it is possible to obtain total bottom coverage over a wide swath beneath the survey vessel. In shallow water it is normal to obtain total sea floor coverage to ensure that no hazards to navigation are missed. A few decades ago a hydrographer would typically gather 1,000 or fewer depth measurements per day. With a modern swath or multi-beam sounding system operating in shallow water, it is possible to obtain more than 1 million depth measurements in a single hour (Fig. 11). In deeper water the number is significantly less but is in excess of 1,000 depth measurements/h.

The data gathering process has not only been enhanced by the introduction of modern acoustic echo sounders. It has also been enhanced by laser mapping systems [11] and by Global Positioning Systems (GPS) that permit continual vessel positioning to a high accuracy. The processing and dissemination of these data still do present many challenges that the hydrographic component of our marine sciences program is attempting to resolve through a number of novel initiatives.

One initiative currently under way as a joint venture with the private sector is the Canadian Ocean Mapping System (COMS). This project involves: (1) the modifying and fitting of a multi beam swath mapping system into a robotic survey vehicle (DOLPHIN — Deep Ocean Logging Platform with Hydrographic Instrumentation and Navigation), (2) the development of a handling system to launch and recover the robotic survey vehicles from a mother ship, (3) the development of a comprehensive survey package to process data, and (4) the transfer of this technology to the private sector. The present status of this project is that the multi-beam sounding system has been fitted into a DOLPHIN vehicle and trials on the handling system are currently under way (Fig. 12). A comprehensive software package to visualize and edit the data is being developed in a cooperative venture between the university community and the private sector and this system is expected to be operational in 1992. It is envisaged that a

Fig. 12. DOLPHIN vehicle steaming abreast of survey vessel

modern hydrographic system consisting of 4–6 DOLPHIN vehicles operating in parallel to a survey vessel will be operational within 5 years. Data will be gathered more quickly than ever before with this powerful survey system, and these data will be machine edited to reduce this time consuming manual task.

With the volume of digital data increasing significantly each year, it is vital that the information be effectively managed and disseminated. A project, underway for several years to implement an effective data base management system (DBMS) to meet hydrographic needs, has resulted in the development of appropriate data structures and techniques that will enable large bathymetric data sets to be managed using relational DBMS technologies [12]. Discussions are currently under way with ORACLE Canada Ltd. to develop this data base management system as a commercial product.

With the ever increasing use of digital techniques for the acquisition of hydrographic data, it follows that navigation charts should be constructed using computer-assisted methods. Chart production has in recent years employed a mix of manual and computer-assisted methods because of the lack of essential information in digital form, lack of computer hardware and software and the lack of skilled personnel. A project is now in place to acquire the hardware and software and to develop the necessary techniques to fully utilize digital techniques for the construction of navigational charts [13]. These initiatives, when integrated with DBMS technologies for data management and artificial intelligence for data generalization, hold the potential for charts to be produced in a much more efficient and timely manner.

While the echo sounder, the radar, and the electronic positioning system have provided major benefits to the mariner, many experts predict that the next technological advancement to significantly impact the community will be the electronic chart. The workload of the officer of the watch will be reduced through the integration of position, depth, and radar information on a chart-like video display and valuable time will be saved when navigating under less than favourable conditions. In addition, the more advanced electronic chart systems have the ability to automatically signal approaching dangers. Canadian hydrographers have been active in electronic chart development through various national and international committees, including participation in the 1988 North Sea Project to demonstrate electronic chart technology [14].

Environmental Responsibility in the Coastal Zone

Population centres and industrial complexes around estuaries have historically discharged wastes into coastal waters — a practice that still continues. Untreated sewage and industrial toxins were believed to be diluted and dispersed by tidal circulation. It is now clear that for many estuarine ecosystems, a legacy of environmental problems has been created that will be difficult, time consuming, and expensive to remedy.

One unforseen factor was the role of sea floor sediments in trapping and accumulating wastes and toxic substances. Recent studies have shown that the history of pollution in some estuaries, with relatively low energy circulation, is remarkably preserved, layer by layer, in geological sequences. It has also been observed that further episodic sea floor processes such as current erosion are able to remobilise wastes from within the sea floor sediments and introduce them to the water column for distribution [15].

Scientists belonging to the Atlantic Geoscience Center of the Geological Survey of Canada at BIO have embarked on a multidisciplinary research project which simultaneously addresses basic science themes and their environmental applications. This project began several years ago, involving marine sedimentologists, stratigraphers, paleo-oceanographers, geochemists, geotechnologists and coastal specialists, addresses a full spectrum of environmental issues including those relevant to health and safety, sustainable development and global change. Some of the specific thrusts of this project are:

1. To develop *in-situ* and laboratory procedures for the determination of *the geotechnical properties of sediments* for stability modelling [16, 17]. Sea floor telecommunications cables, pipelines, wharves, offshore platforms, and defence installations can be damaged by active geological processes involving various types of sediment motion. The purpose of this research is to obtain a better understanding of the mechanics of underwater landsliding in deltaic regions, in potential oil and gas exploitation areas, and in frontier deep water areas.
2. To study *the effect of coastal modifications* caused by large scale engineering and land use changes in drainage basins [18]. The proposed James Bay hydro-electric project in Canada's Hudson Bay (Fig. 13) is an example where river

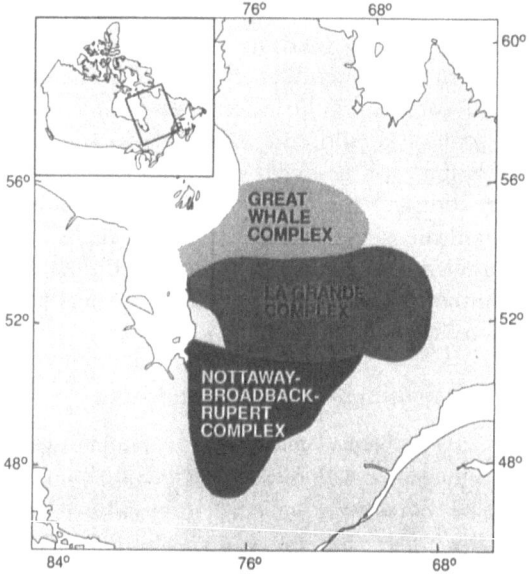

Fig. 13. Hudson Bay

diversion will cause an abandonment of coastal deltas and the creation of new ones with profound effects on sediment supply both locally and regionally and the introduction of exotic and possibly toxic elements to different parts of the marine environment.

3. To investigate *near shore sediments* to obtain information on recent activity of major crustal structures. The long-term design and construction integrity of major coastal facilities such as nuclear power plants is of particular concern where faults are known to have occurred or are suspected. In coastal and lacustrine areas, the near shore sediments can provide a record of the history of recent crustal motion (neotectonics). Mapping of deformation structures and the measurement of geotechnical properties of sediments, both *in-situ* and in the laboratory, allows modelling of motion and the measurement of stresses. This type of information is necessary to interpret regional geophysical patterns and monitored seismic shocks to determine earthquake risk. A project of this nature is underway for part of the floor of Lake Ontario. To date, a microseismic monitoring network has been installed and reconnaissance mapping with high resolution side-scan sonar and multifrequency profilers has been completed. These investigations have revealed diagnostic deformation structures. Forthcoming tasks include systematic surveys to determine the three-dimensional geometry of the lake floor features as well as *in-situ* measurements of stress patterns within the sediments and a reconstruction of deformation styles and a history of the motion that has occurred.

4. To carry out *paleo-oceanographic research* to monitor global change. Marine sediments contain the records of natural and anthropogenic environmental changes necessary to construct global climatic models [19–21]. Patterns of natural changes over geological time scales can be seen as paleo-oceanographic effects. For example, very high resolution analysis of sediment cores off eastern and arctic Canada has shown short-lived (1,000-year time scales) events between 8,000 and 14,000 years ago when ocean conditions changed radically and swiftly. Similarly, research on marine sediments can provide important information on short-term anthropogenic environmental changes such as variations in the levels of carbon and methane gas. Coastal geologists, faced with predictions from climatologists of dramatic changes in relative sea level, are examining coastal sedimentary systems for evidence of accelerating patterns of sea level rise. Over the next decade, coastal responses to changing water levels will also be modelled for various coastline geometries and sediment budgets. Research will also be carried out on the interaction between changing sea conditions and the effects of ice on coastlines. Offshore ice effects on wave energy, coastal ice push processes, permafrost in the near shore and in beach sediments may all be altered by global change. As an important first step, a coastal impact map for Canada is in preparation, using innovative map methods, to display predicted coastal responses to various global change scenarios [22].

5. To develop appropriate technology to efficiently and effectively carry out *geological surveys and research in the coastal and nearshore regions.* The shallow water regions of the coastal zone pose significant challenges to the effective acquisition of geoscience data. It is the highly dynamic nature of the region that is important to monitor and understand, as well as the condition that limits the use of conventional survey technology. Collaborative research with Canadian technology companies into new survey systems, the mooring of equipment, the recovery of seabed samples often with high sand and gravel content, and the requirement for research platforms that are both mobile and stable in shallow water, surf zone areas are the major thrusts in this initiative which will form an important component of environmental marine geoscience research over the next several years. Such challenge can only be achieved through effective networking of scientists and engineers in an interdisciplinary approach.

These studies represent a sampling of the research that is being carried out as part of our environmental responsibility in the coastal zone. The last few decades have witnessed a substantial increase in the input of deleterious materials into coastal waters. This has also been a period of increased public awareness caused by increased media coverage and by a better educated public. The resolution of many of these problems will require an increased scientific effort, interdisciplinary research, and cooperation between the scientific teams and the stakeholders.

Concluding Remarks

What I have attempted to do in this paper is to provide some flavour of the Canadian marine science program with a focus on the research being carried out at BIO, as well as to outline some of the future directions in marine science. We are living in a challenging scientific age with a strong emphasis on large science programs, interdisciplinary research, and numerical modelling. We are approaching an era when environmental responsibility will be frequently tested and an era in which there will be a change from economic regulation to environmental legislation. Close interaction will become more and more necessary between scientists and stakeholders and between scientists and average citizens.

Summarizing from the five future directions that I discussed earlier:

1. To meet the challenge of providing oceanographic information to a growing multidisciplinary scientific community, *analysis tools* are being developed to maximize the information from a variety of data sources. By using techniques to incorporate observational data into statistical and numerical models, information on spatial and temporal scales not previously possible will become a reality. From these observational data and models, oceanographic indices are being created that will be applicable to a broad range of scientific problems.

2. One can anticipate a growing demand from an increasing world population for food from the ocean resource base, a demand that can only be realized by increasing our scientific knowledge in order to *conserve ocean resources and yet increase the yield* from the oceans. Aquaculture is a fast growing industry with considerable potential for increased production. As most of this production will occur in estuaries, inlets and coastal embayments, these features must be studied in a holistic manner in terms of environmental conditions such as minimum temperature, dissolved oxygen availability, and capacity to accept waste products. A multidisciplinary project, involving both field measurements and modelling, is now underway to study one rapidly developing salmonid cage culturing inlet. Through research projects such as this one, it is hoped that the aquaculture industry can continue to develop in an environmentally acceptable manner, staying within the limits imposed by the natural processes.

3. While it has been possible for some time to monitor phytoplankton continuously through chlorophyll measurements, *continuous monitoring of zooplankton* is still a difficult process. Acoustic methods offer a number of advantages over other methods being investigated as they are non-intrusive, provide distributional data in real time and provide high spatial resolution horizontally and vertically. This research thrust which is ongoing is important in terms of obtaining long time-series data on the concentrations of zooplankton and microneckton and also demonstrates the suitability of acoustics in obtaining research information on marine ecosystems.

4. While the gathering of *systematic information on ocean depths* has gone on for many years, it is also a discipline that is undergoing rapid change. Modern multi-transducer and multi-beam survey systems, as well as laser mapping systems, are producing data at a rate orders of magnitude higher than was the case several decades ago. To process these data quickly and efficiently, relational data bases have been developed as well as rigid protocols to facilitate international data exchange. Research on survey systems and vehicles, the development of digital data bases, interactive chart compilation and research on electronic chart technology will provide a fertile field for hydrographic research in the 1990s.

5. *The coastal zone*, the main interface between the land and the marine environment, is the area of highest biological activity and the area of greatest human influence. To protect this important marine region scientists must address the basic science and be fully aware of the environmental consequences of their decisions and recommendations. The resolution of issues in the coastal zone will require interdisciplinary research efforts and close cooperation between the scientific investigators and the users of the coastal zone.

This is but a small sampling of important scientific challenges, opportunities, and problems facing marine scientists in the next 10 years. Some issues not discussed in detail but considered to be important to the ocean sciences, as well as to other fields of research, include the availability of scientific personnel, the question of temporal and spatial scales, the impact of supercomputers, and the attitudes of governments to scientific research. While I have selected a fairly narrow range of future thrusts, I feel, nevertheless, that this is an interesting and challenging era in marine science research and the next 1–2 decades should provide an opportunity for marine scientists to address important scientific issues that will provide significant benefits to the environment and to society.

Acknowledgements. I wish to thank the President of the Japan Marine Science and Technology Center (JAMSTEC) for his invitation to present this paper on the occasion of the 20th Anniversary of JAMSTEC. I also wish to thank Drs. Gordon, Herman, Prior, Ross, Sameoto, and Trites, and Messrs. Burke, Gregory, and Nicholls for either assisting in the preparation or the review of the paper.

References

1. Department of Fisheries and Oceans (1989) Long-term production outlook for the Canadian aquaculture industry. Economic and Commerical Analysis Report No. 13. DFO, Ottawa, Canada
2. Hutchins RW (1990) Canada's oceanic manufacturing and serv ices sector. A background analysis and report to the National Marine Council and the Department of Fisheries and Oceans, March 31, 1990. DFO, Ottawa, Canada

3. United Nations Educational, Scientific and Cultural Organization (UNESCO) (1984) Ocean science for the year 2000. UNESCO, Paris, p 95
4. Petrie B, Drinkwater KF, Pettipas R (1991) Temperature and salinity variability at decadal time scales on the scotian shelf and in the Gulf of Maine. Northwest Atlantic Fisheries Organization Scientific Council Meeting, June 1991, NAFO SCR Doc. 91/86. NAFO, Dartmonth, Canada
5. Silvert WL, Keizer PD, Gordon DC Jr, Duplisea D (1990) Modelling the feeding, growth and metabolism of cultured salmonids. International Council for the Exploration of the Seas (ICES), Mariculture Committee, GM, 1990/F8, Sess. O. ICES, Copenhagen, Denmark
6. Holliday DV, Pieper RE, Kleppel GS (1989) Determination of zooplankton size and distribution with multi-frequency acoustic technology. J Cons Int Explor Mer 46: 52–61
7. Cochrance NA, Sameoto D, Herman AW, Neilson J (1991) Multiple-frequency acoustic backscattering and zooplankton aggregations in the inner Scotian Shelf basins. Can J Fish Aquat Sci 48:340–355
8. Sameoto DD, Jaroszynski LO, Fraser WB (1989) BIONESS, a new design in multiple net samplers. Can J Fish Aquat Sci 37:722–724
9. Herman AW (1988) Simultaneous measurement of zooplankton and light attenuance with a new optical plankton counter. Cont Shelf Res 8:205–221
10. Dietz RS (1970) The underwater landscape. In: Idyll CP (ed) The science of the sea. Thomas Nelson, London, pp 22–41
11. Banic J, Sizgoric S, O'Neil R (1986) Scanning lidar bathymeter for water depth measurement. The Society of Photo-Optical Instrumentation Engineers, Bellington, Wash., USA, pp 187–195 (Laser radar technology and applications, SPIE vol 663)
12. Varma HP, Boudreau H, Prime W (1990) A data structure for Spatio-temporal databases. The International Hydrographic Review (Monaco) LXVII(1):71–92
13. Bell RD, Chapeski RE, Crowther WS, Holman KR, Jackson DM, Orass SR (1989) Nautical chart production using digital data and interactive compilation. Lighthouse, ISSN 0711-5628, edn. 40, pp 25–29
14. Canadian Hydrographic Service Electronic Chart Working Group (1991) The electronic chart in 1991–The CHS changing development role. Lighthouse, ISSN 0711-5628, edn. 43, pp 27–34
15. Paulson AJ, Feely RA, Curl HC Jr, Crecelius EA, Romberg GP (1989) Separate dissolved and particulate trace metal budgets for an estuarine system: An aid for management decisions. Envir Pol 57:317–339
16. Morin PJ, Christian HA, Marsters J (1989) Geotechnical parameters for surficial sediments at the narwhal site, on the south-western grand banks of Newfoundland. Proceedings of the 8th International Conference on Offshore Mechanics and Arctic Engineering, March 19–23, 1989, Genoa, Italy
17. Christian HA, Piper DJW, Armstrong R (1991) Strength and consolidation properties, Flemish Pass: Effects of biological processes. Deep-Sea Res 38(6):663–676
18. Josenhans H, Zevenhuizen J, Veillette J (1991) Baseline marine geological studies off Grande Rivièrede la Baleine and Petite Rivière de la Baliene, southeastern Hudson Bay. In: Current Research, Part E. Geological Survey of Canada, Ottawa, Canada paper 91-1E:347–354
19. Miller AAL, Mudie PJ, Scott DB (1982) Holocene history of Bedford Basin, Nova Scotia: Foraminifera, dinoflagellate, and pollen records. Can J Earth Sci 19(12): 2342–2367

20. Vilks G, Rashild MA (1976) Post-glacial paleo-oceanography of Emerald Basin, Scotian Shelf. Can J Earth Sci 13(9):1256–1267

21. Vilks G, MacClean B, Deonarine B, Currie CG, Moran K (1989) Late Quaternary paleoceanography and sedimentary environments in Hudson Strait. Géographie Physique et Quaternaire 43(2):161–178

22. Shaw J, Forbes DL (1990) Relative sea-level change and coastal response, northeast Newfoundland. J Coastal Res 6(3):641–660 Fort Lauderdale, Florida ISSN 0749-0208

China

FIO Marine Research Programs for the 1990s

Zeshi Chen[1]

Key words. First Institute of Oceanography (FIO) — Research programs — Ocean circulations — Western Tropical Pacific Ocean — Numerical prediction — Disastrous marine environment — Comprehensive investigation — Resources of islands — Ecosystem — Antarctic

Summary. This paper discusses the First Institute of Oceanography (FIO) main marine research programs for the 1990s: (1) ocean observations and research programs for the western Pacific ocean, (2) the research program for numerical prediction of disastrous environments and key techniques in offshore waters, (3) the comprehensive investigation of resources and exemplary exploitation on islands of Shandong Province, and (4) the study on the ecosystem at some key regions (the regions near Great Wall Station and Zhongshan Station) in the Antarctic.

Introduction

China has both land and marine territory, and it is important for the country's economic development to exploit the sea and to develop marine economics. Nowadays, the whole world is facing the problem of population, resources, and environment. One of the fundamental ways to solve these problems is to exploit the oceans.

The First Institute of Oceanography (FIO) has been carrying out scientific and technological activities to facilitate solving these problems, especially since 1980. It took an active part in Government-organized programs such as The National Comprehensive Survey of Coastal Zone and Tidal Flat Resources, Program for Monitoring Marine Environmental Pollution in Adjacent Seas, Numerical Forecasting for the Marine Environment, Marine Surveys for

[1] First Institute of Oceanography, State Oceanic Administration, Qingdao, Shandong, 266003, China

Environment and Resources in the Western Pacific Ocean, Antarctic Scientific Surveys, and so on. FIO has obtained scientific achievements through these activities, resulting in obvious economic and social effects.

In the 1990s, the FIO intends to participate in the advanced activities of the international community of marine sciences and other important international programs. At the same time, it will continue to carry out multidisciplinary research work in marine sciences, which will be combined with basic theoretical research of marine exploitation, monitoring, and management. The research will be aimed at four objectives: national rights and interests, marine resources, marine environment, and disaster reduction. The main research programs are described in detail in the following sections.

Ocean Observations and Research Programs for the Western Pacific Ocean

The FIO is to participate in the WOCE sub-program in China and the TOGA (Tropical Ocean Global Atmosphere)-COARE (Coupled Ocean-Atmosphere Response Experiment) program in the 1990s. FIO will conduct ocean observations in the western Pacific ocean and part of the research work using its experience acquired from implementing the former PRC/US TOGA program.

WOCE Sub-Program in China — Research on Ocean Circulation in the Western Tropical Pacific Ocean

WOCE is a 10-year research program (1990–2000) jointly sponsored by the international organizations ICSU, WMO, IOC, SCOR, and others, which has prompted positive responses from many countries. In August 1989, the Chinese Committee on WOCE was established, thus setting up a solid organizational base for formulating and conducting the WOCE sub-program in China.

The WOCE research program is a continuation of the international TOGA program, and both are important constituents of the World Climate Research Program (WCRP). The program was originated because of the following facts:

- Meteorologists and oceanographers have come to realize that long-term climate changes are largely dependent on global ocean behaviour, and one of the major scientific problems restricting the capability of forecasting these changes is the current inability to describe and simulate global ocean circulation.
- The development of satellite remote sensing techniques, which allow overview observation of global oceans, and the development of large-capacity computers have made it possible to observe, analyze, and simulate global ocean circulation.
- Most scientists in the world community of oceanography have realized that as well as conducting research on the ocean processes at a small or medium scale,

we should simultaneously perform research on ocean phenomena at a global and/or long-term scale.

Ocean circulation in the western tropical Pacific ocean is one of the major components of the global ocean circulation system, and its complexity is known widely. There is the westward south equatorial current, the eastward north equatorial countercurrent, the equatorial undercurrent, the northward Kuroshio source current, the southward Mindanao undercurrent, and the northwestward Papua New Guinea coastal current; all of these make up a very complicated circulation system. These currents play important roles in the redistribution of heat in the oceans and hence in global climate changes as well.

To be able to clarify the mechanisms affecting climate changes with scales of several months to several years would not only speed up the development of ocean dynamics at the levels of air-sea exchange, air-sea boundary layer coupling, and large-scale air-sea interactions, but would also deepen our understanding of the characteristics of the circulation system and its thermal structure in the western tropical Pacific ocean, and thus our knowledge of the mechanisms governing variations in marine environment in the waters over the continental shelf of China. Numerical models for circulation in the western tropical Pacific ocean and for other ocean circulation could be developed on such a foundation and could be used as a basis for climate prediction.

Scientific Aims

The international WOCE Program has two aims: (1) to develop ocean models applicable to forecasting climate change and to collect the data needed for verification of these models and (2) to determine how representative a specific WOCE data set is to long-term oceanic variability, and to explore possible methods for defining the long-term variations of ocean circulations. In agreement with these, the WOCE Sub-Program in China will put emphasis on regional investigations in the western tropical Pacific ocean and research on various mechanisms, so as to achieve the aim of conducting numerical climate prediction at a 10-year level with high resolution over eastern Asia.

Scientific Tasks

In the WOCE Sub-Program in China, the emphasis is to be put on a research into ocean behaviour having especially important effect on climate change over eastern Asia. This research is one of the key links for studying global circulation and the following specific regional tasks need to be carried out:

- Determination of the annual and interannual variations in heat quantities, momentums, and fresh water flux in the western tropical Pacific ocean, Kuroshio region, and South China Sea
- Determination of ocean circulation in the waters of the western tropical Pacific ocean, Kuroshio, South China Sea, and soon, and the response to interface flux, by the method of in-situ measurement and the numerical method

- Determination of oceanic variations with scales of several months to several years in the waters of the western tropical Pacific ocean, Kuroshio, South China Sea, and so on, as well as the parameters and statistical data reflecting such variation
- Investigation of the formation of water masses in the western tropical Pacific ocean, Kuroshio region, South China Sea, and so on, and various processes within the water masses, as well as research into the influence of these processes on meteorological change
- Determination of which ocean variables need to be measured in order to monitor long-term climate change and implementation of technical observation of these variables
- Establishment and development of ocean circulation models for the western tropical Pacific ocean and establishment of a global atmosphere-ocean-sea ice model coupled with the global atmospheric circulation model

Numerical Simulation Research

Numerical Models. China is going to dedicate her efforts to the WOCE numerical simulation research in the following three ocean circulation models:

- Circulation model with a higher eddy resolution. Research on the simulation of regional circulation and corresponding mechanisms is to be performed, with the aim of developing an ocean circulation model with a higher eddy resolution which can be used to define meso-scale ocean processes and those interacting with macro-scale circulations. The first step is to build an ocean circulation model applicable to the regions of marginal seas and the northwest Pacific ocean, and then efforts should be made to achieve an extension of application to global oceans.
- Circulation model with a lower eddy resolution. In the course of using this model, eddies can be produced in principle after parameterization. At present, this kind of model can offer unique results in the field of global ocean-atmosphere coupling.
- Simple ocean circulation model. This model will continuously play an important part in the development of global circulation models owing to the fact that it can result in further understanding of the ocean phenomena and improvement in the relevant models.

Inversion Model and Data Assimilation. The aim of this work is to develop diagnostic models, of which the inversion model will be the major development. Based on data obtained from the WOCE program, a diagnostic analysis is to be performed over the investigation regions in the western Pacific ocean so as to deepen the understanding of the properties of circulation and water masses, and lay the foundation for further improvements to ocean circulation models.

Moreover, it is necessary to conduct the research on techniques of data assimilation in order to establish reasonable initial conditions based on the

WOCE data. Accurate initial values are of great importance to the numerical prediction.

Sensitivity Test. The aim of this test is to find out the effect of ocean circulation on the cycling of some important climate-sensitive material and the effect of various air-sea flux changes on the atmospheric circulation. For instance, a simulation test of the global climate change due to an increase in CO_2 can be performed by using the coupled air-sea circulation model, and the same kind of test can be done on the response to the variation of flux through the air-sea interface (such as heat quantity, momentum, material, and fresh water) by using the ocean circulation model.

Research Contents and Projects

As a marine country located on the west coast of the Pacific ocean, China will put an emphasis on experimental research into the following three ocean components which are of great significance to the ocean circulations: the Equatorial current system, Kuroshio, and its interaction with the circulation over the continental shelf of China.

It is well known that the turning points of various branches of the equatorial current system are located in the western tropical Pacific ocean, or that some of the branches are formed there. The warm pool centered on the section 165° E is a uniquely important phenomenon in the global tropical oceans. Research on the major ocean phenomena, air-sea response processes in this region, and effects on long-term climate change are the key points of the basic research for the sea region.

The Kuroshio originates from the waters of the western tropical Pacific ocean and it is an important channel for longitudinally transporting the heat, momentum, and matter. Moreover, it has an important influence on circulation over the continental shelves in the South China Sea and East China Sea. Meso-scale phenomena, such as meanders, eddies, and undulations, play a key role in the interaction within the Kuroshio and its extension. Research on these problems is of great significance both to the regional oceanography and the parameterization of ocean circulation models.

Some technical problems in the analysis of satellite remote sensing data (e.g., extraction techniques for sea height information) and some basic problems in the numerical simulation, such as the parameterization and the design of highly-stabilized economical computation format, are also to be included in the basic research.

The major research projects are listed as follows:

Physical Oceanography

• Research on the exchanges of energy, momentum, and matter between the western tropical Pacific ocean circulation system and the southern as well as the northern hemisphere

- Research on the formation of general circulation in the Kuroshio source region and the mechanisms of its preservation
- Research on the interaction of circulation in the Yellow Sea, East China Sea and the South China Sea with the Kuroshio
- Research on the volume and location of water mass on the ventilation time scale of 10–100 years
- Research on the dynamics of ocean — sea ice — atmosphere coupling
- Research on the rules of variation in the western tropical Pacific ocean and that of the content of CO_2 in the atmosphere, and their influence on the global climate change
- Research on the temperature-salinity structure of the warm pool in the western tropical Pacific ocean and its relationship with ENSO (El Niño and the Southern Oscillation)
- Research on the variation and transport balance of the current system in the Philippines Basin
- Research on the dynamics of various meso-scale phenomena in the western tropical Pacific ocean

Satellite Remote Sensing Research

- Research on infra-red and microwave remote sensing of seasonal and inter-annual changes of sea surface temperature field in the western tropical Pacific ocean
- Research on the remote sensing of seasonal and interannual changes of sea surface wind field in the western tropical Pacific ocean
- Research on the remote sensing of seasonal and interannual changes of sea height and circulation in the western tropical Pacific ocean
- Research on the remote sensing of meso-scale phenomena in the western tropical Pacific ocean

Marine Chemistry and Trace Element Research

- Research on the interannual changes of nutrient salts in the waters of the western tropical Pacific ocean
- Research on the distribution law of carbon parameters in the waters of the western tropical Pacific ocean
- Research on the change of CFC concentration in the waters of the western tropical Pacific ocean and its application to physical oceanography
- Research on the characteristics and variations of water masses and circulation in the western tropical Pacific ocean based on the distribution and variation of trace elements and by using the BOX model
- Research on the variation of circulation in the waters over the continental shelf of China by using isotope tracing techniques

TOGA-COARE Program

Since 1978, FIO has successively participated in the First GARP Global Experiment (FGGE) (conducting ocean observations in the western equatorial Pacific ocean), the Sino-American Cooperative Research on Air-Sea Interaction in the Western Tropical Pacific Ocean (PRC/US TOGA), Australia Monsoon Experiment (AMEX), and the Equatorial Monsoon Experiment (EMEX), and has made a series of scientific and research achievements. Through the implementation of the PRC/US TOGA program in the last 5 years, the understanding of the following two points has been greatly deepened: (1) the warm pool in the western tropical Pacific ocean is a key water for the occurrence of the El Niño climate event on a short-term scale, and (2) the physical process of the coupled ocean-atmosphere response over the warm pool region is a key step to realizing the mechanism governing climate change on short-term scales. Based on the common understanding of the above two points and according to the requirements of the TOGA-COARE program, an agreement has been reached again between China and the United States, in which a series of *in-situ* observations and research set in the TOGA-COARE program will be conducted co-operatively during the intensified observation period from 1992 to 1993.

Scientific Significance

Among the tropical oceans in the Pacific, Atlantic and the Indian ocean, the western tropical Pacific ocean exhibits unique features. In this region, the southern and northern equatorial currents flow to the west coast and then constitute a complicated circulation system with the upstream of the Kuroshio, Mindanao current, northern equatorial countercurrent, equatorial undercurrent in the subsurface, and the Papua New Guinea coastal current. There, warm water over the widest region and of the deepest depth in the world accumulates all year round. Water above 28° C in this region accounts for 35%–45% of the total in tropical oceans, and its area accounts for 20% of the total surface area of the earth. These unique oceanographic features tend to have obvious interannual variations owing to the fact that these features are governed by various air-sea coupling processes. These variations not only affect climate change in eastern Asia and the western Pacific ocean, but also closely relate to global climate change.

It is well known that the temporal and spatial variations of the warm pool play important roles in the occurrence and development of ENSO. However, the present understanding of the air-sea coupling process in the warm pool region is not enough. For instance, accurate estimation of heat, momentum, and water vapour flux in the air-sea boundary layer, the mechanisms of occurrence and development of the ENSO process, such as the problems of instability theory (Philander) and pseudo-periodic theory (Hirst), and the action of atmospheric compulsory force still remain to be discussed.

Preliminary scientific evidence has shown that the variation in the western tropical Pacific ocean not only relates closely to the occurrence and development of ENSO, but also exerts an influence on the intensity of the Hadley circulation, the variation of high pressure in the subtropical belt, and on the change in frequency of typhoons, through abnormal heating of the atmosphere by the ocean and the variation of northward heat transport of the upstream of the Kuroshio. These variations further affect precipitation and climate in China during the flood season and the sea state in the waters over the continental shelf of China. Therefore, the implementation of the TOGA-COARE program will directly benefit the prediction of climate and sea states in China and raise her adaptability to abnormal climate conditions and sea states, in addition to its contribution to the said international program.

Observation Items

- Atmospheric boundary layer and atmosphere observation by high-altitude balloon sounding, observations of water vapour, momentum, flux of sensible heat and latent heat, radiation flux, and turbulent flow. To conduct these observations, low-altitude temperature profile recorders and Doppler radar atmospheric profile recorders are to be used.
- Ocean observation including current measurement with XBTs (expendable bathy thermograph), CTDs (donductivity-temperature-depth profiles), drifters, ADCPs (acoustic Doppler current profiles), and anchored buoys.

Cooperative Research Contents

- Research on the air-sea coupling process, including air-sea exchange processes for various physical parameters in the warm pool region
- Research on the diurnal variations of various physical parameters in the air-sea exchange process and the interaction between these parameters and the meso-scale climate process
- Research on the west atmospheric convection process
- Research on the interannual variation of the cross-equatorial atmospheric circulation and its relationship with the circulation situation
- Research on the response of the temperature-salinity structure to the meso-scale climate process
- Research on the formation of the equatorial trapped wave and its propagation law
- Research on temporal variations of the upper mixing layer and thermocline
- Research on the responses of sea currents in the warm pool region, the mixing layer, and the temperature-salinity structure to the wind field
- Research on the heat advection process in the upper layer of the ocean

Cooperative Research Waters and Observation Point

- The cooperative research waters are located at 18°20′ N–10° S, 120° E–165° E.
- The observation point is to be set at 2° S, 155° E. Continuous observation at this fixed point for the TOGA-COARE program is to be conducted for 120 days in winter.

- Sectional observations can also be conducted in the course of navigation of research ships to and from the destination according to the relevant requirements.

Way of Cooperation and Period of Cooperative Investigation

Cooperation will be similar to that for the former PRC/US TOGA program. It is stipulated that the Chinese side shall send off an ocean-going research ship to the cooperative research waters to conduct the fixed point continuous observations, and the American side shall offer the necessary investigation instruments and facilities. An allied investigation team is to be organized and the data obtained shall be shared by both sides. The investigation will be carried out between November 1992 and February 1993.

Research Program for Numerical Prediction of Disastrous Marine Environments and Key Techniques in Offshore Waters

Based on the numerical research for prediction of waves, sea temperatures, storm surges, and thermoclines done in the 1980s, FIO will research some subjects under the national research program for numerical prediction of disastrous marine environments and key techniques in offshore waters.

Generally speaking, disastrous marine environments include billows with significant wave heights of over 3.5 m driven by cyclones and typhoons, storm surges, heavy sea ice, anomalous sea temperatures tsunami and so on, and strong winds and tropical cyclones during high sea operations as well. The damage to property and casualties caused by natural disasters has become a severe problem. China suffers heavily from maritime disasters, and the annual economic losses caused by for example, storm surges, billows, sea ice, strong winds, and fog, and tsunami, amount to a few billion yuan, with an economic loss of 960 million yuan caused by storm surge in 1986 alone.

Scientific Goals

The scientific goal of the project is to develop methods to predict disastrous marine environments that are suitable for different stations in the network, and to establish an operational system of numerical prediction. At the first stage, the existing numerical prediction model for marine environments will be improved to give higher accuracy and enhanced ability to allow large-scale model-making central predictions. At the second stage, based on the above model, small-scale models for local areas will be developed so that local marine forecasting centers can make more timely and accurate predictions to form a network numerical prediction system.

Research Objectives

The research objectives of the project are:

- To develop a real-time data receiving and transmitting system
- To develop a real-time monitoring system using satellite remote sensing
- To develop a numerical prediction model for disastrous sea conditions and a tracking and warning system for special disastrous sea conditions
- To perfect the operational systems for the central forecasting station and regional marine forecasting centers
- To develop prediction techniques and operations for marine disasters with emphasis on storm surge, billow, sea ice, and so on. To enhance the effectiveness and accuracy of marine disaster prediction.

Project Tasks

Development of the Operational Systems

The main tasks are:

- To study the assembly and coordination of numerical prediction models for marine environments, establish an operational prediction system for oceanographic data processing and analysis, atmospheric and oceanic models, make comparative tests of similar models to choose the optimal model, and study the composite analyses of sea property fields from different resources and the automatic processing methods for initial field
- To study the objective test and assessment method for marine environment numerical prediction
- To study the interpretation and application of marine environment numerical prediction products.

Satellite Remote Sensing and Monitoring of Disastrous Marine Environments

The main tasks are satellite data collecting, processing and remote-sensing product distribution systems, satellite monitoring of typhoon waves, storm surges, and heavy sea ice, and assimilation of real-time data on disastrous sea conditions.

Study and Operation of Numerical Prediction Model for the Disastrous Marine Environments

The main tasks are the formulation of numerical prediction models for billow, storm surge, and sea ice, consummation of operational numerical prediction systems for the marine environment, establishment of an automatic system for prediction operations, and development of a network system for numerical prediction.

The Operational Forecasting System at Regional Centers

The main tasks are formulation of the numerical prediction models for billow, storm surge, and sea ice at regional centers and the establishment of an operational numerical prediction system for disastrous marine environments at regional centers to make more direct contribution to the prevention of and fight against disasters.

Augmentation of a Database

The main tasks are seriation, normalization, and equality control of historical marine environment data, and the establishment of a GTS surveying database, a predicting product database, and a historic observation and objective analysis database.

Remote Sensing and Measuring System

The main tasks are development and establishment of a microwave/laser remote-sensing and measuring system, development of a phased-array Doppler wind-profiling acoustic radar and acoustic Doppler current profiler, and development of the infrared/laser temperature prefile rediometer. The systems will be used to measure sea surface wind fields, current fields, and sea temperature fields.

Systems Engineering and Strategy Studies

The main tasks are a systems engineering study of the disastrous marine environment numerical prediction, a technical study of testing the level of numerical prediction products for disastrous sea conditions, an information management system for prevention of marine environment disasters, and a study of the influence of oceanic events on climate.

Key Techniques in Offshore Waters

The main tasks are study of a warning and forecasting system for oil pollution in important offshore areas, study of analysis and prediction techniques for anomalous sea temperatures, and study of strong thermocline numerical prediction and analysis of internal waves and fronts.

Comprehensive Investigation of Resources and Experimental Development of the Islands of Shandong Province

In the 1990s, FIO will be in charge of a comprehensive investigation of resources on about 100 islands along the coast of Shandong Province on the basis of our previous work "The National Comprehensive Survey of Coastal Zone and Tidal Flat Resources."

Shandong Province is bounded by the Huanghai Sea (Yellow Sea) to the south and east and the Bohai Sea to the north. From current statistical data, there are more than 200 islands with an area of over $500\,m^2$, but most of these are less than $1\,km^2$. The total area of all islands is $170\,km^2$ and the length of their coast line is about $600\,km$. They are situated near the mainland and form many groups with favorable location, beautiful scenery, and pleasant weather. These islands and the nearby waters have many natural resources and are popular tourist destinations. There is a large potential for development and so, it has become a strategic policy in the development of the national economy to take the islands as bases, develop the sea, and develop the economy vigorously.

To promote the development of islands in the whole nation, the National Science and Technology Committee with another four ministries and national bureaus, issued an announcement launching a comprehensive investigation of resources and experimental development on islands in the whole nation. The purpose of this investigation is to obtain basic data on essential environmental factors, to ascertain the type, quantity, and quality of resources, to check on the social economic condition of the islands with jurisdiction at town level, and to provide a scientific basis for promoting development of the islands. In the past 2 years, every coastal province, autonomous region and municipality directly under the central government has established organizations, worked out plans, collected data and information, raised funds, and carried out the investigation and begun experimental development in stages.

Tasks

According to "Specifications for a National Comprehensive Investigation of Resources on the Islands" and the actual situation in Shandong Province, the main task of this investigation is to investigate and gain data on essential natural environmental factors in over 200 islands in Shandong Province to ascertain the type, quantity, and quality of resources, to find out the present situation of social economic development, and to provide with the data collection maps, reports of the investigation and studies. In addition, research will be done on a program for exploitation and utilization, management regulations, and a policy for development.

On the islands with jurisdiction above the town level, the investigation includes a survey of the natural environment, natural resources, and social economical status of the island itself and its nearby waters. Combined with experimental development of these islands in recent years and the lack of inhabitants, the investigation will focus on the main natural environmental factors and resources. On the other islands there will be only a general survey.

At the same time, a program for island resource development will be required, to choose some islands at the town level as experimental development models, to find out ways to develop the island economy, and to transfer experience to others.

Area

The area of investigation covers more than 200 islands located between the northwest coast of Shandong Province and the south of Haizhou Bay. They will be surveyed region by region. The survey consists of two parts, that on an island itself and that in nearby waters, and the observation will be done using designed stations.

The area to be observed includes the whole island and waters from the high tide line to the isobath of 20–30 m with stations every 5 nautical miles. For the islands to be surveyed generally, observation will be only on the land of the island.

Content and Items

According to the specifications of the national islands investigation, there will be a comprehensive and multidiscipline survey for each island above town level, a special aspects survey for each island worth prompt development, and a general survey for the others.

Format and Content

- The comprehensive survey will include overall investigation of the natural environment, natural resources, and social economy. The content will consist of meteorology, marine hydrology, marine chemistry, forestry, marine biology, environment quality, present status of land utilization, geology, geomorphology and quaternary geology, pedology, vegetation, remote sensing, and social economics.
- An investigation with special aspects will focus on a few selected items. For the islands being developed mainly for beach and shallow water mariculture, the key items are marine biology, marine chemistry, environment quality, and hydrology. For those developing fishery product processing, the key items are social economy, fresh water resources, and fishery resources. For developing a basic construction, they are geology, geomorphology, hydrology, meteorology, vegetation, and fresh water resource.
- The general investigation will survey only the location, area, material components vegetation, and population.

Investigation Items

- The meteorology survey will include following basic meteorological factors: temperature, precipitation, evaporation, wind, atmospheric pressure, humidity, fog, solar irradiance, frostless seasons, visibility, solar energy, and wind energy, as well as weather such as typhoons, cold waves, cyclones, frost, heavy fog, storms, storm rainfall, thunderstorms and hailstones.
- The marine hydrology survey will cover water temperature, salinity, water color, transparency, waves, tides, wind tides, and currents.
- The marine chemistry survey will test for DO, PH, SiO_3^{2-}, PO_4^{3-}, NO_3^-, and NH_4.
- The marine biology survey will survey plankton, benthos, tidal marine life, nekton, primary productivity, and micro organisms.
- The geology, geomorphology and quaternary geology survey will be divided into three parts. The geology survey will include the regional geology of stratigraphy, geotectonics, and mineralogy, as well as hydrology and engineering geology. The geomorphology survey will consist of the morphology, dis-

tribution, erosion, and sedimentation of the land, sea shore and seafloor, the change of the sea shore line, and tourism resources. The quaternary geology survey will study the thickness of sediments, lithology, geological structure, and Quaternary mineralogy and resources.

- The environmental quality assessment will consist of the investigation of pollution sources, pollution discharge of ports and ships, contamination of soil and surface water, and pollution of the littoral zone and shallow water.
- The project will also include investigation of forestry, pedology, vegetation and present status of land utilization, and investigation of social economics.

Duration and Arrangement

This investigation will be conducted from 1990 to 1992 region by region. Some items are carried out seasonally in spring, summer, autumn, and winter. Some are carried out twice a year in the dry season and the rainy season.

Study on the Ecosystem at Key Regions in the Antarctic

In the 1990s, the emphasis of the Chinese National Antarctic Research Expedition will be placed on the problems of resources and environment, and the main study fields will be the following: the ecosystem, living resource potential and exploitation prospects of the Antarctic continent and the southern ocean; crystal structure, formation, evolution, and potential mineral resources in the Antarctic continent and its shelf basin; circumfluence and climate of the Antarctic continent and the southern ocean and their effects on the global climate and environment; Antarctic upper air physics and interactions between the upper layer and lower layer, such as energy coupling and its effects on the electromagnetic wave propagation; and Antarctic medicine and hygiene research.

The scientific activities of FIO in the Antarctic will be concentrated on ecosystem studies at the littoral zone and shallow water near Great Wall Station and Zhongshan Station.

Study Contents

The content of our research will consist of the structure, main function, and dynamics of the ecosystem, with a focus on the area near Great Wall Station. At the same time, the research, will be carried out in the area near Zhongshan Station, to compare these two areas.

Topics

The project will comprise six topics.

Structure of the Littoral Ecosystem

The topic covers species composition and variation of littoral marine life; spatial and temporal distribution of the marine life; interrelations between organisms in the community and the relationship between organisms and the abiotic environment; and phenology of the community.

Function of the Littoral Ecosystem

The topic covers energy pathways of the ecosystem and energy flux efficiency at lower trophic levels; the nutrients cycle, its turnover rate, and decomposition; and biological and ecological restriction, including the restriction of environmental factors and other organisms.

Structure of the Shallow Water Ecosystem

The species composition, abundance, and seasonal variation of living organisms are to be studied, including species composition and variation of the flora and fauna in shallow water; the spatial and temporal distribution of shallow water organisms; the interrelation between marine life, and relationship between organisms and the abiotic environment; and the phenology of the community.

Function of the Shallow Water Ecosystem

The topic covers energy pathways and energy transfer efficiency at lower trophic levels; the nutrients cycle, its turnover rate, and decomposition; biological and ecological restraint, including restraint of environmental factors and other organisms; and the attempt to understand the signal communication in some important population.

Dynamics of the Ecosystem

The topic covers spatial and temporal variation of environmental conditions; seasonal fluctuation of organisms; serial succession of the community; individual development of some species; and the effects of light, temperature, salinity, dissolved oxygen, nutrients, and pH on organisms.

Protection Regulations for the Ecosystem

The regulations will allow clarification of the characteristics of distribution and variation of marine mammals. This will offer a measure of protection for the ecosystem.

Duration and Arrangement

The study will be carried out over the period 1991–1995, and will be arranged as follow:

- 1991/1992. Field sampling and observation, with emphasis on the ecosystem structure at the littoral zone near Great Wall Station, and survey of the environment factors in the region near Zhongshan Station.
- 1992/1993. Continued field sampling and observation. Study on the function and characteristics of the ecosystem near Great Wall Station and Zhongshan Station.

- 1993–1995. Sorting, identifying the samples, data analysis and processing, and research report writing.

Future Developments

In the middle and late 1990s FIO will also participate in the national research projects "The Comprehensive Exploitation and Managemant Program for Bohai Sea and its Bays and Estuaries" and "The Exploration and Exploitation Program of Oceanic Multi-Metal Nodules." Details of these two projects are now being worked out.

Acknowledgments. The author is obliged to Dr. Pu Shuzhen, Dr. Li Bocheng, and Dr. Zhu Mingyuan for offering material on the relevant problems and help in the preparation of this paper.

France

IFREMER and the Scientific and Technological Challenges of the Ocean

PIERRE PAPON[1]

Key words. Exploitation — Marine resources — Coastal oceanography — Deep sea environment — Under sea technology — Climate — Research — Technology — Future

Summary. L'Institut Français de Recherche pour l'Exploitation de la Mer (IFREMER) is a public mission oriented agency. It supports in-house research in marine science and technology and in other French scientific institutions, it operates the French oceanographic fleet (ships and submarines), it is the official expert of the government for fishing resources and the salubrity of coastal waters, and it is engaged in technology transfer to industry.

The Institute's policy is strongly dependent upon its scientific, industrial and social environment: it has to respond to important scientific and technological challenges, it has to contribute to the solution of problems which the society considers to be crucial for its future. This has motivated IFREMER to publish a strategic plan for the years 1991–95, which gives us the guidelines and the priorities for the coming years and in some areas for the decade. The main lines of this plan are presented here, putting a more specific emphasis on the most prospective aspects in four priority areas: the valorization of marine resources, oceanographic research linked to climate prevision, coastal research, and submarine technology.

The exploitation of marine living resources is an important economic activity. Fishing is limited by the availability of resources, and in Europe by quotas, and aquaculture is strongly dependent upon the quality of coastal waters. The main trend of our strategy consists of trying to valorize to the utmost substances extracted from sea products (fish, weed, bacteria), either for their nutritive value or for their chemical and pharmaceutical applications. In a more distant future, we can imagine the chemical exploitation of bacteria recovered close to hydrothermal sources in the Pacific Ocean by our submarine the Nautile, or the development in aquaculture of new varities of fish produced by gene transfer. These are the most challenging aspects of marine biotechnology.

---

[1] IFREMER, 155, rue Jean Jacques Rousseau, 92130 Issy-les-Moulineaux, France

Research on the evolution of the climate is a most difficult question. It supposes studies of oceanic currents, as in the international WOCE program, and of exchanges of heat and matter between ocean and atmosphere, the understanding of biogeochemical phenomena at the surface and in the deep of the oceans. These phenomena play a role in the carbon cycle which is the core of IGBP programs. IFREMER with other French partners (CNRS, universities, CNES, ORSTOM) and in close cooperation with foreign institutions will be strongly involved in such programs which necessitate in situ measurements (with ships and buoys) and global measurements with oceanographic satellites. In the long term, if we are willing to perform climatic previsions, we should be able to organize at the international level the operational observation of oceans with specific instruments (vessels, sensors on satellites, and so on).

Coastal waters, which cover areas including the continental shelf, are particularly important and vulnerable. Coastal ecosystems fix carbon, they are submitted to fluxes of nutrients coming from the continent (phosphorus and nitrogen) and in some areas to fluxes of chemical pollutants (heavy metals, organic compounds). In some coastal zones in Europe, the conjuction of climatic factors with physical and chemical conditions due to continental activities bring toxic algal proliferations (*Dinophysis*, *Alexandrium minutum*). All these phenomena and the necessity of understanding the evolution of the coasts under the influence of currents which transfer sediments motivated a large national effort in France which was launched in 1991, the "National Program on Coastal Oceanography" organized jointly by IFREMER and other French agencies.

Lastly, undersea technology is one of our most important priority areas. It involves research and technological development of new instrumentation, based on acoustics, research on undersea robotics (for teleoperation from a submarine), and on new autonomous and unmanned submarines for the future. The programs are developed in cooperation with industrial partners and in some cases in partnership with European countries.

Oceanographic research is and will be more and more international. Cooperation between IFREMER and Japan has been very active and productive through programs such as KAIKO and STARMER. European countries are actively developing their cooperation in marine science and technology through various schemes, such as bilateral projects or the multilateral programs of the European Community and of Eureka. The prospects of building European oceanographic vessel for high precision oceanic and submarine intervention the "Nereis Project" is one of the most ambitious European projects for the decade.

Oceanographic research is one of the most challenging areas of scientific research as it deals with problems related to resources, environment, origin of life, and evolution of the climate. Its development depends strongly upon the possibility of establishing well-balanced international cooperation.

L'Institut Français de Recherche pour l'Exploitation de la Mer (IFREMER) is a public mission oriented agency founded in 1984 by the merger of two former institutes, ISTPM (Institut Scientifique et Technique des Pêches Maritimes) and

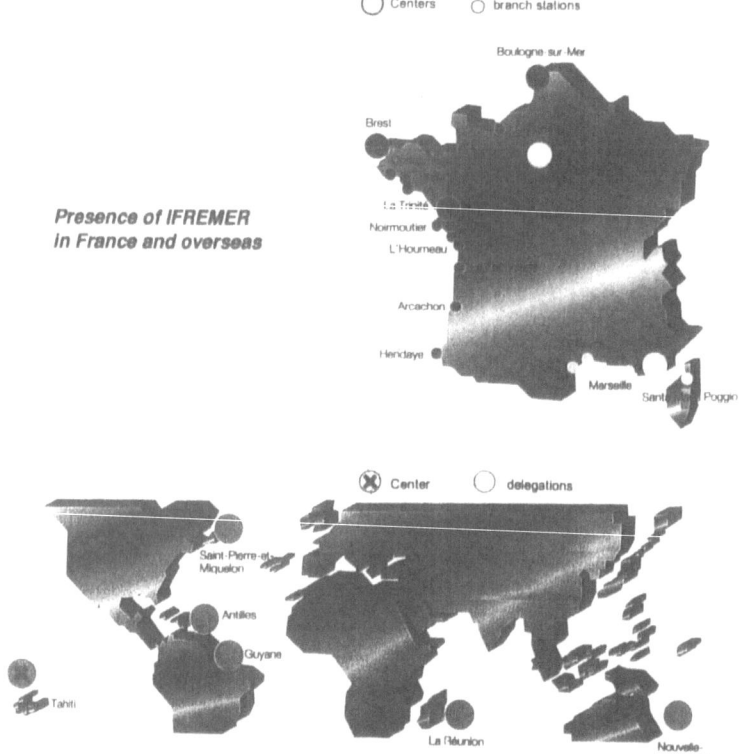

Fig. 1. Localization of IFREMER's laboratories

CNEXO (Centre National pour l'Exploitation des Océans). It is the only French research institution with an exclusively maritime vocation. It supports in-house research and cooperates with other agencies and academic laboratories in marine science and technology. It operates the French oceanographic fleet (ships, submarines, and instrumentation), is the official expert of the State for fishing resources management and the salubrity of coastal waters including cultured shell fish, and is engaged in technology transfer to industry.

Supported by the Ministries for Research and Technology and of the Sea, its laboratories are located on the French coast. IFREMER has a centre in Tahiti (French Territory of Polynesia) and a research station in New Caledonia. The Pacific ocean having always been considered by IFREMER as one of its most important and active "laboratories" (Fig. 1).

The present budget is FF1 billion and there is a staff of 1,800 (scientists, engineers, clerks, technicians, sailors), some of whom are employees in subsidiary companies such as GENAVIR, our fleet operator. IFREMER has an oceanographic fleet with nine ships and two manned submarines, and associated equipment.

The Institute's policy is strongly dependent upon its scientific, industrial, and social environment. It has to respond to important scientific and technological challenges involving the ocean and to contribute to the solution of problems which the society considers crucial to its future [1]. This means that we have to fix priorities to concentrate our scientific and technological potential on the most promising areas in coordination with French marine policy. This has motivated IFREMER in October 1991 to publish a strategic plan for the years 1991–1995, which gives the priorities and the guidelines for our scientific and technological policy for the coming years and in some areas for the decade [2].

I will present the main lines of this plan here with emphasis on the most prospective aspects in four priority areas: the valorization of marine resources including the development of marine biotechnology, research on coastal waters, oceanographic research, and underwater and naval technology.

Research and Technology for the Exploitation of Marine Resources

The exploitation of marine living resources is an important economic activity in the world. Global catches of fish and aquaculture production, both sea and freshwater, amount to roughly 100 million tonnes worldwide. In Europe, decisions regarding fishing resources and their exploitation within the EC Exclusive Economic Zone (EEZ), as well as sanitary regulations for aquaculture are taken at the level of the European Community on the basis of scientific advice prepared by a major scientific organisation, the International Council for the Exploration of the Sea (ICES). This is the so-called Blue Europe policy.

During the last decade, the productivity of the fishing vessels has been increasing continuously while the availability of resources remains limited. The recognition of this fact has guided and will guide in the future the two main priorities of our research policy in the area of fishing research: explaining the natural variability of the resources in relation to climatic and hydrological factors, and enhancing the selectivity of fishing.

One can think of fishing as a fish-fishermen system; the variation of fish stocks depends upon such parameters as fishing techniques (nets, trawlers), power and characteristics of fishing boats, and so on. However, such a systematic approach, which was performed in the 1970s, is not sufficient. Biological, climatic, and physical factors must also be taken into account to explain the fluctuations of fish populations. The impact of the climatic event El Niño on the anchovy fisheries in Peru is thus very well known.

The critical phenomenon requiring study is the so-called recruitment of fish, which refers to the population of young surviving fish becoming exploitable every year. To understand its annual variation, the influence of biological and physical parameters on its very existence is crucial to determine a reasonable fish catch from year to year. Research in this area will be active during this decade, involving disciplines such as biology of development, population dynamics, ecology, and chemical and physical oceanography.

Selectivity of Fishing Vessels and Fishing Instruments

A second aspect of fishing research, in France as well as in other countries, is related to technology and covers the development of techniques for the detection of fish, the design of new kinds of hull, storage facilities and fishing nets for vessels, and so on. In the future, this activity, which is important within IFREMER, will be guided by a priority: increasing the selectivity of fishing vessels and instruments. This has been motivated by the necessity of catching only fish within the allowed size range and of the highest commercial value. Technological research aims firstly at improving the evaluation of stocks and the identification of species through acoustic means. A multi-beam acoustic detector is thus being tested within the framework of the HALIOS project (Eureka program), which can discriminate between the sea bottom and the fish and possibly between different species of fish. Another direction is the design of new trawlers and the simulation of their functioning modes. This involves computer-assisted design and testing of highly-innovative hull forms through the use of models in hydrodynamic basins such as the one inaugurated in 1991 in Boulogne. Other studies have been initiated for trawlers in order to reduce fuel consumption of the boats.

Sea Farming

Sea farming has become an important source of protein: world production amounts to roughly 10 million tonnes of fish, shellfish, molluscs and crustaceans, seaweed, and other aquaculture products. The development of this activity in Europe has certainly been sustained by an increasing demand for fish protein: fish, and more generally sea products, are forming a larger part of the diets in European countries. The important French tradition of oyster farming has motivated a strong IFREMER involvement in research in this area. One of our main efforts in the future will be devoted towards obtaining species able to resist specific diseases (marteiliose and bonamiose), which supposes the development of good research capabilities in molecular genetics and biology.

Another area of excellence has been and still is research on tropical sea farming. Intensive prawn breeding and raising techniques developed by the IFREMER Pacific Oceanological Center in Tahiti and our station in New Caledonia for the constitution of genitor stocks, coupled to a specific feed defined by IFREMER, have confirmed that prawn species, such as *Penaeus vannamei*, *Penaeus monodon*, and *Penaeus stylirostris*, are well suited to tropical waters. Physiology of reproduction and development, nutrition, and the capacity for disease resistance will be our research priorities in the future.

Sea farming of salmon and trout has been very successful in Europe (particularly in Norway) and we have devoted considerable effort with the Institut National de la Recherche Agronomique (INRA) to develop this activity in France. During this decade, priority will be given to species such as turbot, bass, and sea bream. This will put an emphasis on the breeding production parameters of larvae physiology, nutrition, and development.

Finally, we have recently turned our attention to algae farming. Macrophytous algae culture aims at the production of algae for use as food and to provide industry with the vegetable matter it needs. Our research programs have the main objective of mastering the reproductive cycle and seed production in species such as *Porphyra linearis*, *Laminaria digitata*, and *Laminaria japonica*, and cultivation is being investigated on the Mediterranean coast.

Valorization of Sea Products

Our research strategy has been guided until now by the recognition that the ocean is considered a field for mass production of living species exploited either by fishing or by sea farming. Looking to the prospects for the decade, one can draw guidelines for a new strategy, which of course in many areas are related to our present programs. The exploitation of living resources should give particular priority to qualitative aspects. Thus the biological limitation of traditional fish stocks available for fishing encourages us either to discover new species not caught at present or to invest in research to be able to systematically exploit fisheries' by-products. This is termed the valorization of sea products, and it implies, for example, conceiving new processes for surimi production from fish protein and the extraction of peptide fractions through chromatographic methods from industrial hydrolysates of fish proteins. These peptides are potentially interesting for the pharmaceutical industry as they have immunostimulant properties. Algae are another area of interest. For example, some species of macrophyte algae have biological activity, as in the case of sulfate polysaccharides extracted from brown algae.

This research strategy relies on the progressive emergence of marine biotechnology, involving such important areas as genetic engineering, molecular biology, and immunology as applied to marine species. These species live in specific physical and chemical conditions: salty water, the presence of various chemical compounds that sometimes include heavy metals and sulphur compounds, the presence or absence of light, and heavy pressure in deep ocean. They emit chemical signals for communication, defence, and attack and they have specific means for the metabolization of various compounds. Scientific research must take into account these peculiarities in marine species to use their metabolites or "genetic engine" to produce chemicals and pharmaceuticals with various properties (immunostimulants, hypotensors, antitumoral drugs, neurodrugs).

This is the objective we are pursuing at our laboratory of marine biotechnology in Brest through, for example, work on bacterial strains collected close to hydrothermal vents in the Pacific ocean by our submarine Nautile. Some of these thermophilic bacteria live in temperature conditions close to 100° C, which means that they have specific enzymes and polysaccharides able to survive and work at such high temperatures. They also have the capability of metabolizing sulphur compounds. Research on these thermophilic bacteria and their potential applications is one of our priorities.

Naturally, one must not forget the potential applications of genetic engineering to marine species, such as the transfer of genes with interesting specific properties from one species to another: to produce bacteria that manufacture new enzymes, and fish or oysters resistant to disease or capable of faster growth. The potential applications of genetics and molecular biology are so important that IFREMER has decided to create a new laboratory in Montpellier in order to develop this research area. It will be a cooperative laboratory with scientists from CNRS (Centre National de la Recherche Scientifique) and the University of Montpellier.

Offshore Sea Farming

The exploitation of marine living resources, is now strongly dependent upon environmental conditions. Climatic factors do have an influence on fisheries, as well as the quality of coastal waters on sea farming and coastal fisheries. This is the motivation behind our decision to focus technical research efforts on so-called offshore sea farming, i.e., the development of offshore platforms and related equipment to exploit sea farms some miles offshore. This would be particularly important along the Mediterranean shore where there is insufficient coast with good biological and physical conditions. This also explains the increasing coupling between research areas in marine resources and the coastal environment.

A High Priority: Coastal Oceanography

Coastal waters, which cover areas including the continental shelf, are particularly important and vulnerable: coastal ecosystems fix carbon, they are submitted to fluxes of nutrients coming from the continent (phosphorus and nitrogen) and in some areas fluxes of chemical pollutants (heavy metals and organic compounds) [3]. In some coastal zones in Europe the conjunction of climatic factors with physical and chemical conditions due to continental activities (agriculture, industry, urban development) bring toxic plankton proliferation (*Dynophysis*, *Alexandrium minutum*, a producer of paralyzing toxins). All these phenomena and the necessity of understanding the physical evolution of coasts under the influence of currents and of coastal equipment (wharfs, harbours, etc.) motivate the important research effort which we are undertaking in this area and which is IFREMER's main priority for the decade.

The excess of continental nutrients in coastal waters causes eutrophication, a proliferation of microplankton (coloured waters) and macroalgae (green waters). Eutrophication is particularly prevalent on the coasts of Britanny and is quite a nuisance. IFREMER has studied the various physical and biological parameters which explain eutrophication and developed a mathematical model which is able to describe it with fair accuracy. Explaining the periodic development of toxic algae is more difficult. Further research is necessary for an understanding of the mechanism of its apparition (for example, by developing methods for in vitro cultivation of toxic plankton) and the origin of the toxins. This will enable us to conceive systems of detection and possibly of prevention.

Research on Urban Waste Treatment

Urban waste treatment has, of course, a great impact on coastal waters. This has prompted the urban wastes program set up last year by IFREMER. Die-off processes were studied in bacteria such as *Escherichia coli* and *Salmonella*, and results showed the importance of coastal water quality on the process. In oligotrophic Mediterranean waters, bacteria die quickly. By contrast, in the Atlantic Ocean and the English Channel, sediments and suspended matter improve the survival of bacteria because dissolved and particulate matter contains osmoprotectors and nutrients which can be assimilated. Moreover, bactericidal solar radiation is attenuated by suspended matter. Research has shown too, that the elimination of nitrogen and of phosphorus by water treatment plants is generally weak. Applications of this program are now being persued in coordination with public agencies in charge of water management.

Research on urban wastes, and an understanding of the functioning of coastal ecosystems or of eutrophication phenomena cannot be restricted to purely biological or chemical approaches. It is important to rely on mathematical modelling of currents and their mixing based on oceanographic data. Several digital models have been developed to describe mechanisms at different scales. The development of a reliable model of phenomena using a set of biological, chemical, and physical data will be very important for the future. It will allow us to predict the impact of coastal equipment on water quality (sewage channels, harbors, sea farms), and as such, will be an effective tool for the management of coastal zones. Uncontrolled spills of organic chemicals (pesticides for example), heavy metals (mercury, cadmium) and nutrients in the sea represent a growing risk for the salubrity of coastal waters and thus for sea farming. Our research effort is necessary to alert public authorities and to develop expertise.

The National Program on Coastal Oceanography

In the near future we must concentrate research activities on three main questions:

1. What will be the long term effects of the continuous enrichment of coastal waters by nutrients and organic matter?
2. Will some chemicals, even at very low concentrations, present serious risks for some marine species? (Will they be able to modify their genomes?)
3. Shall we be able to guarantee the quality of water to maintain sea farming activity?

To answer all these questions we shall need to sustain an important program involving diverse disciplines such as ecotoxicology, marine chemistry, biology and genetics, and hydrodynamics. It also requires a research effort in instrumentation (innovation in sensors for example) to monitor the waters continuously. Therefore, in 1991 we launched a National Programme on Coastal Oceanography to federate the efforts of IFREMER, CNRS, marine stations, and the French ministries having an interest in the area. Furthermore, the necessity of innovation in the field of instrumentation has led us to organize a parallel effort with industrial partners to develop automatic systems (buoys) for the collection of physical and chemical parameters in coastal waters and data transmission through satellites. In the long term, this program should completely modernize our system of surveying the quality of the marine environment. The project will provide a national network for the observation of oceans (detection of pollutants, bacteria such as *Salmonella*) supported jointly by IFREMER and the French Ministry of Environment, and it constitutes an important mission for us.

Remote Sensing: An Important New Tool

Remote sensing will certainly play a growing role as a tool for coastal environment management. The Polynesian remote sensing station in Papeete (managed jointly by IFREMER and the Territory) as well as the ORSTOM (French Institute of Scientific Research for Cooperative Development) station in Noumea and the IFREMER Center in Brest have shown the applications of high resolution SPOT images to pollution monitoring and map making for the Pacific island environment. The use of radar images such as those taken by the European Space Agency (ESA) ERS 1 satellite will give us complementary means for operational monitoring of coastal waters.

Oceanography: From Deep Sea Environments to the Climate

We used to say in France that the ocean is a vast laboratory covering the whole planet. As it would be tedious to describe all experiments taking place in such a laboratory, I prefer to but emphasis on the areas which seem to us the most promising.

Plate tectonics has been a very active research area. Our efforts have been mainly devoted to the study of oceanic accretion in dorsal basins (Fiji Islands, for example) and dorsal zones in the Atlantic ocean within the framework of major bilateral projects such as STARMER (in cooperation with Japan) and FARA (in cooperation with the United States) or multilateral projects (INTER-RIDGE).

Research on subduction phenomena has been the centre of some very fruitful cooperation with Japan in the Kaiko project. A French team of scientists, headed by Prof. Xavier Le Pichon, and their Japanese colleagues, have used surface vessels and the submersible Nautile (Fig. 2) to investigate the Nankai Margin quite extensively. It is an area which is the seat of considerable deformations and where mechanical energy is dissipated in earthquakes. With a basic understanding of the structure and nature of this margin it is now possible to identify with fair confidence the main faults and landslides that will probably be activated above the potential earthquake focus. Furthermore, massive zones of active seeps of fluids have been detected on the sea floor. Thus, marine geophysics gives us the capacity of using instruments on the sea floor to monitor tectonic evolution until an earthquake does occur. This is the motivation behind Prof. Le Pichon's proposal to permanently monitor and repeatedly survey the sea-convered portion

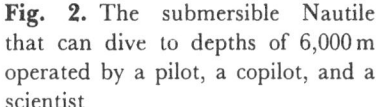

Fig. 2. The submersible Nautile that can dive to depths of 6,000 m operated by a pilot, a copilot, and a scientist

of the Tokai earthquake area to detect geophysical events which could be advance signals of an earthquake.

Hydrothermalism: A Research Priority

IFREMER, in cooperation with other French laboratories, has been widely involved in research activities in this field that take into account all facets of the phenomena: physical, chemical, and biological. The project resulted in the discovery in 1990, within the context of a back-arch (Lau Basin), of the most active site known in a classical dorsal ridge (Franco-German Nautilau Cruise). In future, we will devote our efforts towards understanding the mechanisms of hydrothermalism, the thermal and material balance of the hydrothermal fluids (in some areas they are enriched with metallic sulphurs and flow at 350°C) and their contribution to the build up of mineral beds on the bottom of the ocean.

The study of hydrothermal ecosystems is of course a very challenging subject to which IFREMER has devoted significant efforts. It is known that animal colonies (bacteria, bivalves, vestimentiferan worms) live in the near vicinity of hydrothermal vents at medium and high temperatures. Metallic sulphur and other elements can be recovered from hydrothermal fluids in animal tissues. All

Fig. 3. L'Atalante, the new IFREMER oceanographic vessel commissioned in October 1990. It is a multi-purpose research vessel with 450 m^2 of scientific premises. Equipped with the EM-12 multibeam echo-sounder, she is capable of operating a manned submersible

ecosystems draw their energy from hydrothermal sources in which hydrogen sulphur and methane are present.

Our research strategy in this area is twofold. Firstly, we want to investigate the ecological system that these animal colonies constitute, particularly with the role of associated chemosynthetic bacteria. Hydrothermal vents constitute a most remarkable environment which has not changed during the 3.7 billion years since the beginning of the primitive ocean; archaebacteria still exist in hydrothermal vents at 100–120°C and they might be related to the bacteria of the primitive ocean. This is quite a challenging perspective for specialists: what is the place and role of these archaebacteria in the evolution of living species? Secondly, as mentioned previously, we study hydrothermal ecosystems from the promising point of view of their potential technological applications. More than 100 bacterial strains, now undergoing close analysis, are growing at temperatures above 80°C. Three directions are explored: the study of membrane lipids, the study of bacterial exopolymers such as polysaccharides, and research on thermostable enzymes. Another possibility is to study the specific heat-stable collagens from the giant vestimentiferan worms, such as *Riftia pachyptila*, living close to hydrothermal vents which help them to sustain life at high temperatures and pressures.

An Important Priority: The Study of Climate Evolution [4]

Let us just recall the crucial role that oceans play in the evolution of climate. They take part in the exchange of energy and matter between earth and the atmosphere, acting as a kind of thermal engine. They have a large capacity for carbon dioxide absorption, thereby buffering the so-called greenhouse effect which has to be evaluated. This explains why IFREMER, with other French partners including CNRS, academic laboratories and ORSTOM, has concentrated its efforts on taking an active role in main international programs such as WOCE (World Ocean Circulation Experiment), TOGA (Tropical Ocean Global Atmosphere), and JGOFS (Joint Global Ocean Flux Study). We contribute both scientists and oceanographic vessels, such as L'Atalante (Fig. 3), to these multidisciplinary programs that involve physical oceanographers, biogeochemists, and specialists in hydrodynamics. To study the sea's deep layers in the WOCE program, IFREMER has commissioned the development of a type of floater which is followed acoustically and is capable of surfacing periodically to transmit its data using the ARGOS system to allow for the reconstitution of their trajectories.

The ocean cannot be studied only by surface or subsurface means, ships and buoys. We need global observation of phenomena and satellite data measurement. The launching of Europe's first earth remote sensing satellite, ERS 1, in July 1991 was a turning point for oceanography. ESA's satellite now transmits altimetric data, including the height of waves and the overall height of the surface, wind direction and speed, and surface temperature. The synthetic aperture radar (SAR) on ERS 1 produces images of excellent quality along the track of the satellite and particularly at high latitudes. It records a swathe of the surface, 100 km across. IFREMER, through CERSAT in Brest, archives and treats the satellite data except for the SAR images. The considerable advantage provided by the ERS 1 has led us to launch two programmes, one specifically related to the exchanges between ocean and atmosphere, the other focused on icy seas. Both of them make extensive use of satellite data and are closely related to the climate studies. They are expected to engender intensive international cooperation.

Technology: A Challenge and a Tool

Engineering and technology is a very active research area within IFREMER and our strategy is guided by the need to develop specific instrumentation for the scientific community, to support technological developments useful for French industry, and to develop challenging new systems, particularly for underwater activity. Present and future plans focus on automatization of most systems as a high priority.

One of the best examples of our achievements in instrumentation which illustrates this idea is the Nadia diagraphy shuttle which offers the possibility of reintroducing measuring devices in holes previously drilled during Ocean Drilling Program (ODP) cruises (Fig. 4). The shuttle is operated by the Nautile, to which it is coupled. It can carry out thermal measurements, fluid sample collections, magnetic mesurements, and acoustic imagery of the inner walls of wells without calling upon a specialized drilling ship. The tests have been satisfactory.

For the future, instrumentation for deep sea activities as well as for the coastal environment will be developed. In both cases, discovery of new sensors is crucial. It might be worthwhile to investigate biosensors for their capacity to detect traces of pollutants and optical devices to detect objects and transmit information. Furthermore, we perform research in two directions as far as underwater observation in concerned: (1) taking and transmitting images by acoustic means (the TIVA system), and (2) using lasers to perform in situ measurements (light-

Fig. 4. The shuttle Nadia. Operated by the Nautile, the shuttle can introduce measuring devices into previously-drilled holes

scattering techniques, for example) and to realize detection of objects in the sea and recovery of optical images. All this is undertaken in close cooperation with CNRS and academic laboratories.

Offshore Structures and Naval Technology: Two Major Areas Related to Underwater Activity

The improvement of tools for the design and survey of offshore structures that have to withstand strong pressure from swell, currents and winds, requires good field knowledge, statistical representation of data for structural calculations, and study of their behaviour. In this field, perform collaborative research projects with IFP (Institut Français du Pétrole) and engineering companies. In the future, calculation and dimensioning methods for platforms designed for great depths (more than 500 m) will certainly be a most challenging area. In parallel, operations at sea, such as the transfer and installation of platforms and dimensioning of constructions along the coast and in the high seas, depend on the

Fig. 5. The NES 24, an air-cushioned catamaran, which will start tests in 1992

climatology of the sites concerned. The feasibility of satellite observation of sea conditions has been demonstrated by IFREMER for high latitudes in seas with icebergs, such as the Hibernia oil field close to Newfoundland.

Naval hydrodynamics is a fairly active research area in France, within our own laboratories in colleges of engineering (in Nantes for example) and in the Defence establishments. Our specific interests are twofold: solution of problems occurring during the study of new ships and development of calculation tools with the support of appropriate experimental means in order to contribute to the advancement of knowledge concerning the stability and behavior of ships. We are engaged in a naval technology composite materials program concerning fishing vessels and passenger ships with various French partners. A technological venture for the future with great potential is the MENTOR programme (Modèle Exploratoire de Navires de Transport Océaniques Rapides) which was launched in France in 1991. It is dedicated firstly to the identification of the most promising system for a high speed (50 knots) passenger ship (1,000 tonnes) with a large autonomy (500 nautical miles) and high reliability. Secondly, it will undertake specific technological research to solve problems of materials, naval architecture, navigation, and so on. IFREMER has a fair experience with surface-effect ships through the NES 24 project (an air-cushioned catamaran) which will be completed in 1992 (Fig. 5), but we want to investigate other techniques such as the SWATH system (Small Waterplane Area Twin Hull). The MENTOR project will be undertaken in cooperation with all the French intitutes having a competence in naval technology including IRCN (Institut de Recherche de la Construction Navale), colleges of engineering, Defence naval research institutes, and shipyards.

To complete the engineering activities in relationship with industry, I should mention that ocean mining engineering is a potentially important area. The discovery of polymetallic nodule fields on the bottom of oceans world-wide and more particularly in the Pacific ocean is an asset. The exploitation of the polymetallic nodule ore deposits has been considered as a possible source of

Fig. 6. The SAGA industrial submersible built by **IFREMER** and **COMEX**. Equipped with a hyperbaric compartment for six divers and able to dive to 600 m, the Stirling engines gives it a large range

metals, principally nickel, manganese, cobalt, and copper. France has been granted a mining licence for the future exploitation of these deposits in the Pacific ocean within the framework of the International Law of the Sea. IFREMER, in cooperation with the French Commissariat à l'Energie Atomique (CEA) and industrial partners such as Minemet and Preussag in Germany, has performed research in chemical engineering to compare processes proposed for the treatment of deep sea nodules (the sulphuric leach process ranks first) and has carried out a project to develop a prototype for an automatic vehicle to harvest the nodules underwater. Due to the suspension by most industrial groups of any investment in ocean mining, the French government has decided to maintain only a technological follow up activity in the coming years.

Present and Future Projects for Submarine Vehicles

As well as the two manned submarines (the Nautile which can dive to depths of 6,000 m and the Cyana which can dive to 3,000 m), IFREMER can operate the SAGA (sousmarin d'assistance à grande autonomie or high autonomy assistance submarine). This submarine, (28 m long, 550 tons of displacement), has been developed by IFREMER and Compagnie Maritime d'Expertise (COMEX) and carried out its first tests at sea in 1990. It is the first prototype of a new generation of industrial submarines. It is equipped with hyperbaric compartments for six divers, can dive to 600 m and its Stirling engines allow great range. It established a new record in May 1990, releasing its divers at 316 m for more than 4 hours (Fig. 6).

Necessity of Developing New Kinds of Submarine Vehicles

Observing the bottom of the ocean, taking fluid or rock samples as well as bacteria, and investigating submerged shipwrecks are examples of functions which such a system is expected to be able to carry out under the sea. All these operations suppose accurate sea floor positioning of the vehicles, and we are upgrading our methods in this area using acoustical techniques.

A further possibility is the use of manned submarines such as IFREMER's Nautile coupled to automatic devices. The Nadia shuttle is an example of such a solution. This technology requires development of advanced robotization with computer-assisted remote manipulation as a central technique. We have developed a hydraulic "slave" arm specially conditioned for underwater applications. In the same spirit, the inspection robot ROBIN tethered to the Nautile increases the capacities for investigation of shipwrecks. Future developments may include manned submarines with large range operating remote-operated vehicles (ROV) which can be repaired or retooled on board under the sea.

There are also completely different solutions possible with unmanned systems and vehicles. Benthic laboratories for sea bottom observation are one example of such automatic devices. They will be designed for both short- and long-term observations and in situ measurements (of a hydrothermal field for example) in the framework of the Nereis project. Cables for power and fiber optic links from sea floor to surface would connect it to an oceanographic vessel. The benthic laboratories would have to be located on the sea floor by acoustic means and controlled using a video camera, and they would carry scientific instruments and sensors. ROVs for scientific operations with three degrees of freedom and a capability for observation and manipulation either on metallic structures or rocks and hydrothermal vents are certainly interesting tools. Some oceanographic institutes are already equipped with such systems and the development of an ROV for deep sea operation (6,000 m) is being planned by IFREMER.

Last but not least, the development of a completely autonomous underwater vehicle is one of our high priorities. This vehicle could have a large range (at least 3 days undersea at a speed of 10 knots) with a vast operational capacity. We have chosen to develop an automatic vehicle dedicated to cartography and bathymetry. Such a project requires the resolution of complex problems such as energy storage, navigation, and propulsion. We shall be engaged in the first phase of development of the system in active cooperation with the British Natural Environment Research Council (NERC).

Underwater technology is a very challenging area. It is the key for future progress in deep sea oceanography and ecology.

Looking to the Future

Exploration and exploitation of the ocean is most important for our society. The oceanic world remains largely to be discovered and understood, and this raises important scientific questions and requires development of high technology.

Firstly, going back to the problems of the exploitation of resources, we can consider them on different time scales. In the short term (the decade), understanding the best conditions for exploitation of marine resources by fishing and sea farming in conjunction with environmental conditions is most urgent [5]. This has to be done at international level, not forgetting the necessity of cooperation with developing countries for which marine resources are an asset (for example Africa). It is also probable that the exploitation of offshore oil fields at great depth (below 500 m) will develop during the decade and this will stimulate technological developments in areas such as platform technology, underwater robotics, and ROVs [6, 7]. During this decade, important international research projects will probably be launched to study the impact of anthropogenic activies on deep sea environments (embedment of chemicals in sediments, the impact of industrial waste spills on ecosystems, etc.), which is a problem that IFREMER has taken into consideration. Secondly, 10 years from now, offshore industry might require the development of industrial submarines with great range in various sea conditions, such as under the ice pack.

In the longer term (such as 20 years from now), exploitation of polymetallic nodules and frozen gas hydrates such as methane hydrate for example might be possible. This would require specific methods for the exploitation of these resources. Once more, automatization will be a necessity in this area as well as for the ships of the future.

International Cooperation

Oceanographic research will require more and more complex equipment both for oceanographic vessels and submarines. IFREMER is ready to investigate with foreign partners possible solutions for extended international cooperation on building and operating oceanographic vessels. First of all we intend to rely on European cooperation which is developing quite intensively and extensively; the EC program on Marine Science and Technology (MAST) and the Eureka program with several projects related to marine technology (the Halios project between France, Iceland and Spain for the technical development of future trawlers) are effective catalysts for this cooperation. IFREMER has proposed the Nereis project to its European partners. This project consists of building a European ship capable of light drilling (300 m in sediments under 6,000 m of water) and of handling underwater instruments (such as benthic laboratories) located on the ocean floor. The European proposal would use a rather light platform to answer the specific needs of the global change community and extend our capacities for experimentation in the deep sea. This would be a European project operated by a consortium, open to cooperation with non-European partners. The proposal takes into account the necessity of deep sea drilling in oceanography but also that it is a tool so diversified and so complex that it is more and more difficult to run associated programs with a single drilling platform as within the Ocean Drilling Program (ODP). We need to renew completely our ideas and perspectives in such areas; the Japanese and CIS initiatives to build new platforms are positive as they offer new possibilities.

Global Observation of the Ocean

I will conclude by emphasising the necessity, which we shall face more and more, to complement our in situ measurements in oceans with a global view of the ocean surface. This involves, of course, measurements of surface parameters (temperature, wave height, wind speed, presence of plankton). Spatial geodesy using altimetric data provided by satellites (GEOSAT, ERS 1, the Franco-American TOPEX-POSEIDON in the future) will bring progress to the field of marine geophysics (plate tectonics for example). Oceanography performed with satellites is now coming to maturity [8]. Furthermore, we have to persuade governments of the necessity of performing a global observation of the ocean. Obtaining data with spatial and temporal continuity is necessary if we want to understand (and in the very long term to forecast) climate evolution. This will require a worldwide system with a network of satellites, automatic surface and submarine vehicles, vessels, and buoys with data transmission capabilities. To set up such a system is the great challenge of the decade.

We feel that progress in oceanography coupled with the use of instruments on satellites (large platforms or small satellites) will at the turn of the century promote an "operational" oceanography allowing, for example, transmission of useful data for navigation (in Arctic seas particularly), fishing, industrial activities, and coastal monitoring. Oceanography at large could open challenging perspectives for our society if we scientists are able to show our fellow citizens the possibilities of marine resources and the necessity of protecting the marine environment of the planet.

References

1. Gaudin M (1990) 2100 "récit du prochain siècle." Editions Payot, Paris
2. IFREMER Plan stratégique 1991–1995. IFREMER, Issy-les-Moulineaux, France
3. Bourgoin J, Alzieu C (1990) L'environnement littoral. Quels enjeux pour la recherche? IFREMER, Issy-les-Moulineaux, France
4. Le Gal Y, Barbier G (1990) Biotechnologies marines. L'état des recherches en France et à l'étranger, les enjeux, la place et le rôle de l'IFREMER. IFREMER, Issy-les-Moulineaux, France
5. IFREMER (1991) La mer et les rejets urbains — colloque de Bendor (juin 1990). IFREMER, Issy-les-Moulineaux, France
6. IFREMER (1991) Technologies sous-marines pour la recherche et le développement, colloque de Toulon (1990) IFREMER, Issy-les-Moulineaux, France
7. Stone GS, Busch WS (1991) Deep ocean science facility needs marine technology. IFREMER Reportion Society Journal 14–25
8. ERS 1 (1991) ESA Bulletin ERS 1 (special issue): 65

Japan

JAMSTEC: Present and Future

Isao Uchida[1]

Key Words. Utilization of coastal sea — Deep sea survey — Deep sea research — Biological and geological data — Biological environment of the deep sea — Oceanographic observation — Oceanographic research — Cooperation — Future — Deep sea drilling — The Observation and Research Center

Summary. The Japan Marine Science and Technology Center (JAMSTEC) was incorporated in 1971 as a general oceanographic institution, its management coming under the general supervision of the Science and Technology Agency (STA). Besides the Administration Department, it has, for research and development work, the Deep-Sea Research Department, the Deep-Sea Technology Department, the Ocean Research Department, and the Marine Development Research Department. For fleet operations, there is the Ship Operations Department. Since its establishment, JAMSTEC has been actively engaged in basic and pioneer-type R&D projects, shared use of large-scale facilities, educational and training programs, and technical library and information services, thereby contributing to the development of Japan's oceanic science and technology.

Some of the primary achievements of JAMSTEC to date are as follows: On the sea floor of the continental shelf with depths of up to 300 m, the temperature is relatively constant throughout the year, so that it is possible to create a high-pressure environment utilizing the ambient water pressure. This sea floor space on the continental shelf can be effectively utilized if humans can be made to work there directly. JAMSTEC has been engaged in developing this diving technology under the "New Seatopia" Project, and has carried out psychological and physiological studies during dives. For the next step, we aim at developing a comprehensive system for underwater work employing unmanned technology to the maximum degree possible.

[1]Japan Marine Science and Technology Center, 2-15 Natsushima-cho, Yokosuka, 237 Japan

JAMSTEC took note of the energy of waves existing naturally on the sea surface and carried out a series of sea trials for effective use of this energy via the experimental power generation system "KAIMEI." The experiment demonstrated the practicability of developing large wave power generation systems and established a design method for an economical KAIMEI-type wave power generation system. Based on this research experience, JAMSTEC now aims to develop technologies to calm the sea area behind a system designed for efficient wave energy utilization, to help improve the environment of the sea area.

The oceans of the world have an average depth of about 3,800 m. In other words, our earth is covered mostly by those deep oceans and seas. We are thus called to uncover the realities of the deep sea. JAMSTEC has engaged in the development of tools necessary for deep-sea research in forms of manned and unmanned submersible systems. For manned submersibles, we have the "SHINKAI 2000," with a depth capability of up to 2,000 m, and the "SHINKAI 6500," with a depth capability of up to 6,500 m, currently the world's deepest. For unmanned probing, we have the "Dolphin 3K," with a depth capability of up to 3,300 m. All of these submersibles are in active service now, and the discoveries of hydrothermal phenomena in the Okinawa Trough and of active black-smoker spurts in the same area are a few examples that demonstrate their utility. Also, JAMSTEC is engaged in basic studies of deep-sea organisms growing in those environments where such phenomena are observed.

JAMSTEC is also attempting to understand the dynamic processes of the ocean. Our efforts include studies on large-scale oceanic change phenomena and the development of technologies required for carrying out such studies, e.g., ocean acoustic tomography and ocean laser profiling. We are also participating in the joint international programs of TOGA, WOCE, and JGOFS.

With the twenty-first century now in perspective, JAMSTEC's emphasis in the years to come will be on the development of comprehensive research and development programs. We will aim at deep-sea research which will, by utilizing our state-of-the-art equipment such as the SHINKAI 6500, hopefully result in achievements worthy of world recognition. We expect to pursue observation programs which are global and comprehensive. Regarding areas of research and development, we will need to establish priorities taking into consideration the situation anticipated 10 years from now. We will also need to take special measures to give effective support to these programs. For the immediate future, we consider the following appropriate as our priority areas of research: (1) deep-sea surveys and research, (2) ocean observation, and (3) coastal sea area development and utilization. We also consider the following important as our basic approaches to promote our work: (1) establishing long-term and specific research objectives, (2) ocean observations, (3) balance between science and engineering; (4) JAMSTEC functioning more like a "center of excellence" in ocean research and development, (5) strengthening the environment to support research work, and (6) cooperation with other organizations including participation in joint international programs.

Introduction

The Japan Marine Science and Technology Center (JAMSTEC) celebrated its 20th anniversary in October 1991. On this occasion, I would like to overview our activities during these past 20 years and to present our future plans. I would appreciate any advice and comments from those who have participated in this symosium.

Outline of JAMSTEC

JAMSTEC was incorporated in October 1971 by investments from government and industry on the basis of the Law of the Japan Marine Science and Technology Center (JAMSTEC) approved by the 65th Diet.

For several years after its establishment, funds were supplied from government and industry on a 50-50 basis. But as the nature of activities changed, the government contribution kept increasing, and this year it has reached 90%. The total budget for the 1991 fiscal year was ¥ 11.7 billion, or roughly US$80 million. JAMSTEC has carried out advanced research and development in close collaboration with industrial groups, academic groups, and the government. The nature of collaboration has included guidance from the Science and Technology Agency (STA) of the government, advice from scholars and persons of experience on various committees, and joint R&D with other research institutions and private companies.

Concerning JAMSTEC staff, the number of permanent employees is 156 for 1991. However, nearly twice as many people work at JAMSTEC altogether, when the crew of the three vessels and researchers on temporary assignment from other institutions and companies are included.

As the social needs have changed with the times, the scope of research and development activity at JAMSTEC has changed and kept expanding. At present, the research and development activities of JAMSTEC are divided into three major categories: the first is research and development for the utilization of coastal seas; the second is research and development relating to deep seas; and the third, oceanographic observation and research in the open sea (Fig. 1).

Research and Development Activities in JAMSTEC

Research and Development for the Utilization of Coastal Sea

In the early 1970s when JAMSTEC was established, the exploitation of continental shelves attracted the Society's attention. It was thought then that undersea activity by man would be essential for the exploitation of natural resources and for the utilization of space on the continental shelf. Therefore, JAMSTEC at that time started the project "SEATOPIA" in which undersea

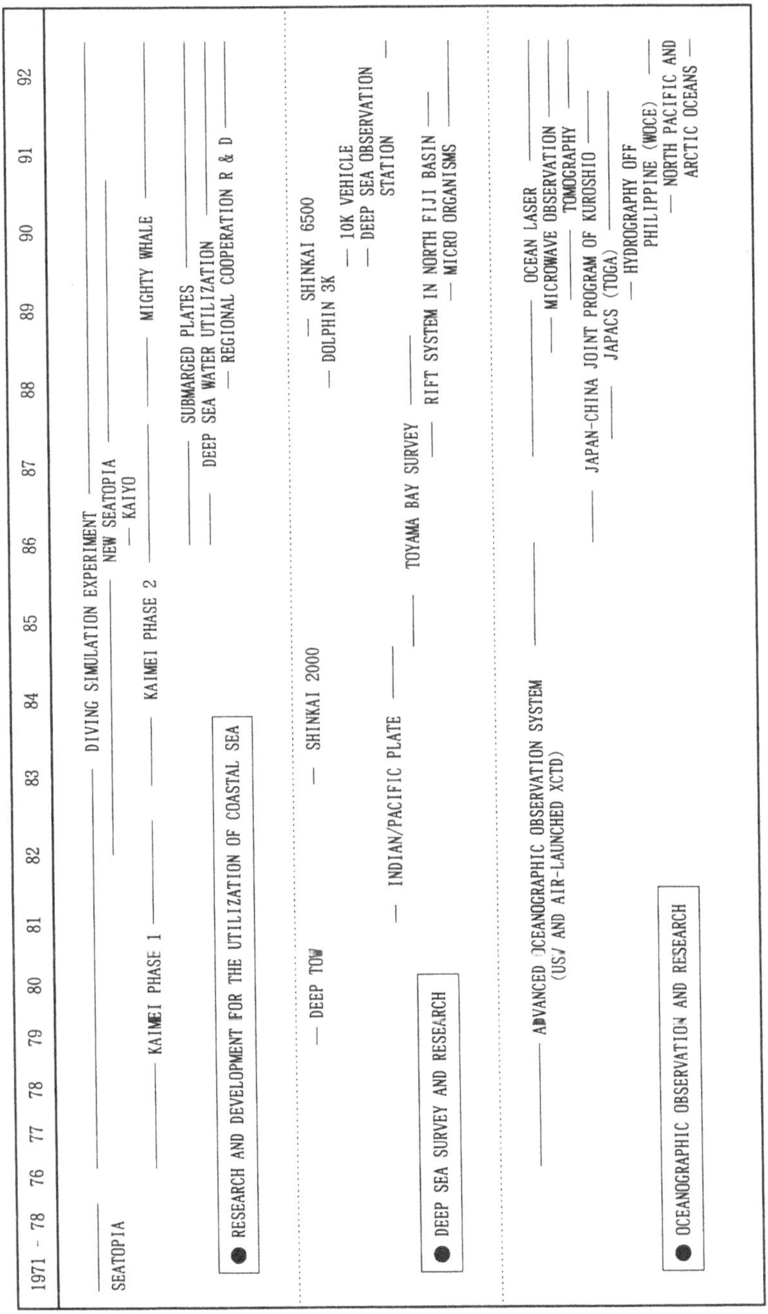

Fig. 1. Project at JAMSTEC from 1971 to present

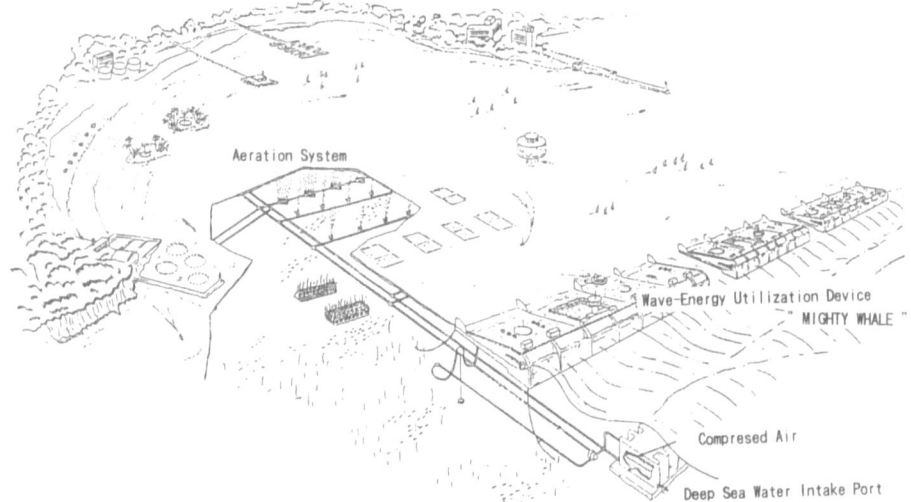

Fig. 2. Offshore floating wave-power system, Mighty Whale

habitation experiments using a saturation diving system at the depth of 60 m were conducted.

In 1985, after verifying the safety of deep-sea diving with the use of a land-based simulation facility, JAMSTEC built "KAIYO," a semi-submerged catamaran type vessel (SWATH) which has a gross tonnage of 2800 t and is suitable for saturation diving experiments. Through the "New SEATOPIA" project using KAIYO, we established in 1990 the technology for saturation diving up to depths of 300 m in the open sea.

We have conducted studies on the physiological and psychological characteristics of diving including animal experiments in cooperation with, among others, the Wisconsin State University and the University of Hawaii, to confirm the safety of diving activities. We have also conducted physiological studies on breath-holding diving fishermen called "ama," in collaboration with the New York State University. With future utilization of wave energy in mind, JAMSTEC started an open-sea test of the wave power generator 'Kaimei" in 1978, as a joint project under the International Energy Agency (IEA), in co-operation with the United States, the United Kingdom, and so on. This project has proved the viability of wave power generation in the ocean. Since 1988, based on the results of the Kaimei project, JAMSTEC has conducted research and development on a new floating wave-energy utilizing system, "Mighty Whale," which can absorb wave power efficiently and leave a calm sea behind the device (Fig. 2).

In order to develop and utilize coastal seas effectively, it is necessary to consider specific local requirements and to conduct joint technology development with appropriate local users. In this context, since 1988, we have developed a number of new technologies for the utilization of coastal seas in close cooperation with local governments. For example, in Kochi prefecture, we have successfully demonstrated that deep-sea water can be used for aquaculture production as it is rich in nutrients, low in temperature, and clean. In Okinawa prefecture, basic research has been conducted to develop technology for coral transplanting in order to restore damaged coral reef communities. In Kumamoto prefecture, we have developed a technology to control passage of fish schools by means of an electric fish screen. Last of all, in Iwate prefecture, we are developing a new artificial sea bed with submerging and surfacing functions for culturing fish and shellfish at optimum depths.

We believe the technology and experience gained through these projects will be useful not only along the coasts of Japan, but also in many other parts of the world, and we would be happy if we could share them with other countries.

Deep Sea Survey and Research

Since we started the development of a deep tow system in 1979 to survey 6,000-m ocean depths, we have developed a variety of deep sea survey systems, such as manned submersibles and remotely operated vehicles, in order to challenge the vast deep sea which remains unknown to us.

The untethered manned submersible "SHINKAI 2000" that is capable of diving to 2,000 m was constructed in 1981 and has been in active service for diving research since 1983. It has participated in survey activities with its mother ship "NATSUSHIMA" mainly around Japan, in areas such as Suruga Bay, Sagami Bay, the Izu-Bonin area, the Nansei Shoto Islands, and the Japan sea, and has made a total of 574 dives as of the end of September 1991. The construction of the untethered manned submersible "SHINKAI 6500," capable of diving to 6,500 m, was completed in 1989 (Fig. 3). It has been engaged in deep sea surveys with the "YOKOSUKA" since early 1991 around the Japan Trench and the Fiji Islands. The YOKOSUKA, the support vessel for SHINKAI 6500, has a gross tonnage of 4,439 t. As with the NATSUSHIMA, YOKOSUKA is used as an independent research vessel and is equipped with a multi-narrow-beam echo-sounder and other advanced instruments.

JAMSTEC has also developed unmanned survey systems for longer durations and for use in wider or dangerous areas, where the use of manned submersibles is not appropriate. Besides the deep tow system mentioned above, JAMSTEC has developed various free swimming vehicles such as "HORNET" which is capable of diving up to 500 m. Another example is "DOLPHIN 3K" which is a large, heavy-duty ROV (Remotely Operated Vehicle) capable of surveying up to depths of 3,300 m. Furthermore, in aiming to be able to survey even the deepest seas of the world, we are manufacturing an ROV with full ocean depth capability, that is, with a capability of diving up to 11,000 m. We expect to complete the ROV by 1993.

JAMSTEC is now engaged in research and development of other newer vehicles. One of them is an optical-fiber-controlled free-swimming vehicle in which the

Fig. 3. Manned research submersible, Shinkai 6500

hydrodynamic drag of ordinary cables has been eliminated. Another one under development for a next gereration system is an autonomous completely cable-free underwater vehicle system.

We have been able to obtain varied biological and geological data including some with high scientific value by making use of these research vehicles. SHINKAI 2000 discovered live *calyptogena* and *Vestimentifera* communities near cold seepages on the sea floor in Sagami Bay. This was the first discovery of such a biological community in the Western Pacific. In the Okinawa Trough, black smokers were discovered and the phenomenon of evaporating clathrate CO_2 from the sea bottom was observed for the first time in the world.

SHINKAI 6500 started its scientific diving service this year and has already discovered some fissures on the sea bottom at 3,100 m on the Okushiri Ridge in the Japan Sea. This suggests that an active faulting process is now in progress in this area. Another discovery worthy of mention was quite fresh cracks 10–15 m wide and 3–5 m deep on a slope of the Japan Trench at a water depth of about 6,200–6,300 m. The cracks are considered to have been derived from the bending of the Pacific Plate. In addition, JAMSTEC has been involved in the STARMER program, a research program on rift systems with the cooperation of France and conducted with financial support, for Japan's part, from the Special Funds for Promotion of Science and Technology of the Japanese Government. This research program has spanned the 5 years from 1987 to 1991, with the aim of studying accretion processes in the North Fiji Basin at a microplate located between Fiji and New Caledonia in the Southwestern Pacific. This year, the last year of the joint program, SHINKAI 6500 conducted dives in this area, and we obtained large quantities of new scientific data. Many research institutions, both domestic and overseas, participated in the program including IFREMER and ORSTOM of France, institutions of South Pacific countries, the Geological Survey of Japan, and so forth. This is a good example to show that the cooperation of many research institutions, both domestic and overseas, is required to execute such a large-scale research program. The results obtained during these 5 years will be presented by the scientists concerned with this program at the International Geological Congress which will be held next year in Kyoto.

Research on the biological environment of the deep sea is also important. It has been proved that there exist organisms depending not on solar energy but on chemosynthetic bacteria utilizing chemically enriched sea water, such as the above-mentioned *Calyptogena* living in hydrothermal and cold seepage areas in the deep sea. JAMSTEC considers the study of these ecosystems found in the special high-pressure environment in deep sea to be among the most important future research subjects.

The purpose of this research is to understand biological, physical, and chemical processes taking place in the special environment of the deep sea. Since 1990 we have organized the Deep-Sea Environment Program which is open to the outside scientific community and invites brilliant researchers from all over the world to conduct advanced basic research. As the first step, JAMSTEC has looked at deep-sea microorganisms and, especially, their unique genes and proteins not found among microorganisms on land. One of the very first results of this program was our recent discovery, among a sample collected by SHINKAI 2000 from the sediment of Suruga Bay, of a microorganism that dissolves petroleum. We are currently developing an experiment system, including a high pressure incubator, which allows us to bring back deep-sea microorganisms and culture them on land under the same condition as at the deep-sea bottom of up to 6,500 m. Not only that, but we are also constructing an all-in deep sea environment research laboratory to enable us to conduct comprehensive research most efficiently.

Oceanographic Observation and Research

There have recently been serious discussions on global environmental changes such as global warming and desertification caused by human activity. Abnormal weather at various places in the world has come to be understood in the context of global climate change. The right understanding of global climate change requires the right understanding of oceanic change. Above all, studies of heat transport and ocean flux in the ocean and their interactions between the ocean and the atmosphere are important. The traditional oceanographic observations, however, have not been performed on a global network basis, like the one used in atmospheric observation, and have thus produced only discrete data sets. In order to observe oceanic changes, we must develop new observation systems using advanced technology. We have been engaged in the development of new ocean observation systems since 1977. First, a towed underwater sliding vehicle

(USV) equipped with CTD sensors was developed to observe the mesoscale eddies of warm core rings off Tohoku, Japan. We succeeded in a transect observation of warm core rings there using the USV in 1988.

In 1987, as the next step, with a view to developing a new system which can achieve wide-range, simultaneous, and three-dimensional monitoring and which can respond to the need of global oceanic observation and research, we started developing microwave radiometry, ocean laser, and ocean acoustic tomography technologies. Aircraft or satellite-borne microwave radiometry, based on the intensity of emissions from all materials, can detect the conditions of water vapor, rain, sea ice, snow, and so on. We are now developing an aircraft-borne system. Ocean laser technology aims to obtain vertical profiles of the volume of phytoplanktons, important data for ocean flux studies in the sea surface layer. The measurement is based on fluorescent and scattering strengths in the water of a transmitted laser pulse. A prototype apparatus for ship use has been made and we are now collecting fundamental data for accurate measurement. In the future, it is planned to develop an aricraft-borne laser (Lidar). Ocean accustic tomography is a new observation technology which can measure the water temperature and the direction and speed of currents, based on the principle that the speed of sound in water depends on the water density. Variations in the propagating time of acoustic pulses are converted into a density profile between a transmitter and a receiver placed far from each other. Furthermore, when many transmitter-receiver sets are deployed, an instrantaneous estimate of three-dimensional density distribution in the ocean can be obtained. Now, we are developing a 1,000-km unit, that will use a new low-frequency transducer using magnetostrictive material. We are developing practical models through open sea tests. It is planned that JAMSTEC and the Woods Hole Oceanographic Institution (WHOI) would conduct cooperative observation off Bermuda in the near future (Fig. 4).

Besides the above-mentioned technologies under development, the following observation projects are under way in the Pacific in cooperation with other domestic and foreign organizations.

The Japan-China cooperative study on Kuroshio started in 1986. JAMSTEC has carried out oceanographic observations, such as the long-time mooring of current meters to study the variability of the Kuroshio current velocity in the shelf slope of the East China Sea, and CTD profiling using underwater sliding vehicles to study mixing processes at the shelf edge.

The Japanese Pacific Climate Study (JAPACS) began in 1987 as a 5-year program. In this program, JAMSTEC has studied heat transport in the ocean mixed layer in the Tropical Pacific through hydrographic and meteorological observations using the research vessel NATSUSHIMA. We have studied ocean variabilities in relation to El Niño, such as current distribution using acoustic Doppler current profilers, and sea surface temperature and surface height using satellite data. In these cruises, the Meteorological Research Institute carried out atmospheric observations including radiosonde sounding and chemical tracing. We realize the importance of obtaining results through collaboration in which

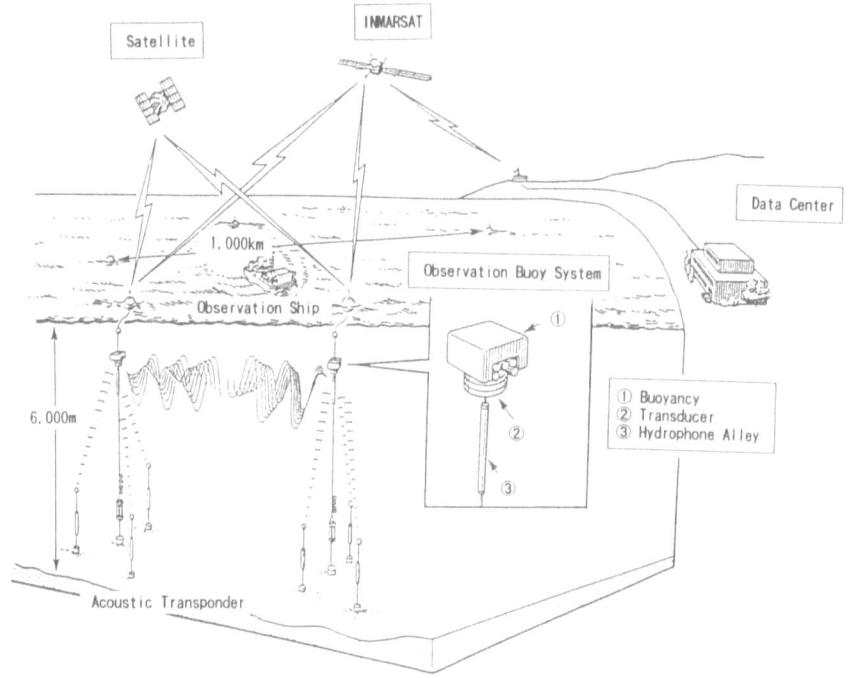

Fig. 4. Ocean acoustic tomography observation system

each participating party conducts observation in its own specialized fields. In order to join the TOGA-COARE (Tropical Ocean Global Atmosphere Program-Coupled Ocean Atmosphere Response Experiment) whose observation exercises concentrate in 1992 and 1993, it has been decided by STA that the 5-year program will be extended for 1 additional year.

JAMSTEC has begun a study on hydrography and mass transports off the east of the Philippines, as part of WOCE (World Ocean Circulation Experiment), focusing on the variabilities of the North Equatorial Current, the Mindanao Current, and the Mindanao Eddy. The first cruise using KAIYO was conducted in 1991 by JAMSTEC.

We began oceanographic research in the northern Pacific and the Arctic oceans in 1991 and carried out hydrographic observation in the Bering Sea. Furthermore, observational study using an Ice-Ocean Enviromental Buoy deployed on polar ice is being conducted with WHOI on ocean circulation and ocean flux. We aim to perform integrated observation including ocean circulation in the northern Pacific and flux study in the Arctic oceans.

A preliminary test for measuring global warming of the ocean was carried out in early 1991 based on the acoustic propagating time between Heard Island in the Indian Ocean and points deployed in each of the oceans of the world. The project, an application of ocean acoustic tomography technology, was proposed

Table 1. International cooperative projects at JAMSTEC (govermental and inter-govermental)

• US-Japan Cooperative Program in Natural Resources (UJNR)	1972–
• Japan-China Joint Program of Kuroshio	1986–92
• Heat Transports in the Western Tropical Pacific (JAPACS/TOGA)	1987–92
• Hydrography and Mass Transports off Philippines (WOCE)	1990–94
• Research Program on the Rift System (STARMER)	1987–91
• Visiting researchers supported by STA fellowship	1989–

by Dr. Munk of the Scripps Institution of Oceanography and we took part in the project to catch the acoustic signal in the South Pacific.

International Cooperation

In addition to those projects mentioned above, we have participated in the following international cooperation efforts.

As a leading Japan-side member organization in the Diving Physiology and Technology Panel of the United States-Japan Cooperative Program in Natural Resources (UJNR), JAMSTEC has conducted joint research on high pressure physiology and deep sea surveys with the National Oceanographic and Atmospheric Administration (NOAA) (Table 1). Since 1988, a number of joint efforts have been conducted with WHOI under an inter-institute agreement including the development of a new-generation autonomous unmanned vehicle. We are signing a new agreement for cooperation with the Scripps Institution of Oceanography, too. Thus we have not only participated in various multinational international projects, but have also conducted research and development jointly with individual foreign institutions (Table 2). There is no question that the results of these cooperative activities have greatly contributed to the advancement of our capabilities. We hope the same has been true with our counterparts.

Cooperation with Domestic Organizations

We feel that our joint research with domestic universities and industry in developing deep-sea submersibles and in surveys utilizing them has represented an important part of R&D activities of JAMSTEC. We would like to continue our R&D efforts effectively and efficiently in cooperation with these organizations.

Future Plans

As I have mentioned, during these past 20 years JAMSTEC has kept expanding its areas of activity from the coastal seas to the deep sea and further to open sea areas, and has been able to achieve results centered on technology. On the basis of this, I would like to express my views on what JAMSTEC is going to do and under what kind of philosophy in the years to come.

Table 2. International cooperative projects at JAMSTEC (inter-institute)

● Cooperative studies with Woods Hole Oceanographic Institution	1986–
● Cooperative studies with Scripps Oceanographic Institution	1986–
● Cooperative study on saturation diving	1989–
● Cooperative experiment for air-launched ocean laser	1990–
● Global ocean variability acoustic measurment	1991–

Table 3. Strategy for the twenty-first century

- Balance between science and technology
- Long-term objectives and systematic approach to them
- Cooperation with domestic and foreign institutions
- Active participation in international programs
- Improvement of the environment for research and development
- Among centers of excellence in the world

On the occasion of this 20th anniversary, JAMSTEC is now working on a long-range plan for the coming decade (Table 3). In that plan, JAMSTEC would like to set guiding principles as follows:

1. *Balance between science and technology.* An important part of oceanographic research is to gain scientific knowledge to understand the nature of the ocean. This can be achieved only through the development of instruments and systems that represent state-of-the-art technology. Close cooperation between scientists and engineers is thus essential to modern oceanographic research. As mentioned earlier, JAMSTEC started with more weight on technology, and on this basis, has expanded its activities into scientific areas. From now on, JAMSTEC, while continuing to develop advanced technology based on experience, will also endeavor to strengthen its scientific research capability, and aims to grow as an institute balanced well between science and technology.

2. *Establishment of long-term targets and systematic programs.* It is not possible to grasp the widespread and complex processes of the ocean within a short period of time. Therefore, we will set up long-term, concrete goals for R&D, develop well-coordinated programs including multidisciplinary projects to achieve these goals most effectively, and attempt to implement these programs according to a timetable.

3. *Reinforcement of ties with domestic and foreign institutions.* In Japan, oceanographic research has been conducted by various expert institutions in their respective fields of interest. For example, the Meteorological Research Institute has accumulated excellent results in the field of oceanic climatology, as has the Fisheries Research Institutes in fisheries research. Universities have also carried out their own academic research. Therefore, in the implementation of broad integrated oceanographic research, multidisciplinary cooperation

among these institutions is indispensable. One of the important roles of JAMSTEC, as stipulated in the Law of JAMSTEC, is to conduct broad integrated research in cooperation with other institutions. As well as broad research programs, multidisciplinary linkage and interchange of researchers are also of extreme importance in individual research projects. We will strive to further strengthen the ties with related institutions. This applies to foreign institutions as well, especially to those which, like the participants in this symposium, are engaged in broad research in ocean science.

4. *Active participation in international programs.* It goes without saying that no research covering a wide ocean area can be implemented effectively by any one institution or country alone. Therefore, we will actively participate in, and cooperate with international programs currently underway or proposed. At the same time, we will be proposing the initiation of some new international programs centering around the Pacific Ocean.

5. *Improvement of the environment for research and development.* In order to carry out effective, integrated research and development, capable researchers and advanced instruments and equipment are indispensable. We will provide for needed resources of both types to improve the environment for research and development.

6. *Centers of excellence in the world.* We hope that JAMSTEC will be able to become one of the centers of excellence in the world. Through the realization of this we aim to serve, some day in certain areas of oceanographic research and development.

Based on the guiding principles mentioned above, we would like to set up long-term targets for R&D activities at JAMSTEC. One of the most urgent issues of the world in the next 10–20 years concerns global environmental change and, to deal with it, it is very important to understand the earth scientifically. Above all, JAMSTEC considers a subject of the highest priority the need to get a fairly clear overall picture of the ocean, which covers about 70% of the earth's surface. JAMSTEC will facilitate establishing large-scale observation network systems to cover the Pacific Ocean, which occupies half of the whole ocean and in which Japan is located. Using these network systems, we will attempt to understand the structure of the area and the processes taking place there, under the best organic relations possible with domestic and foreign institutions.

We also think the utilization of the ocean is essential to the continued well-being of human beings. Therefore we would like to develop advanced ocean utilization systems making use of new knowledge and technologies that will not affect the natural environment and ecosystems irreversibly. On the basis of this fundamental philosophy, our targets for research and development in the coming decade are described below.

Deep Sea Research

Looking worldwide, important subjects of deep sea research include the study of geological structures of the earth and plate movement processes in the deep sea

Table 4. Deep sea research

- Studies of evolution of ocean floor based on plate tectonics theory
- Studies of carbon cycle in deep sea
- Studies of deep sea organisms
- Studies of sub-bottom layers in ocean floor

area useful for prediction of earthquakes and tsunami, the study and utilization of deep-sea living organisms with characteristics totally different from those living on land, and the study of circulations of chemical and biological materials. With these in mind, JAMSTEC would like to set its targets for deep-sea research as follows (Table 4):

1. *Deep-sea crustal structures based on plate tectonic theory.* We would like to conduct research on deformation pocesses of the ocean bottom in the Northeast Japan Transect, the northernmost part of the Philippine Sea, and other plate convergence areas, and research on the plate formation processes in the East Pacific Rise, Mid-Atlantic Ridge, Southwest Pacific, Okinawa Trough, Mariana Trough, and so forth. Thus we will concentrate our efforts in the Pacific Ocean, but we will deal with the Mid-Atlantic Ridge in the context of the Inter-RIDGE Program. Using the results obtained through the studies above, we would like to complete simulation models for deformation of subducting plates, interactions with landside plates, and consequent occurrence of earthquakes.

2. *Sub-bottom structures.* We would like to develop a deep-sea drilling system with a view to studying the geological structures and dynamics of the oceanic crust, with sub-bottom samples being collected.

3. *Circulation of materials.* We would like to undertake research to develop a comprehensive understanding of the processes using the seas around Japan as a model, including transport and degeneration of life- and terrigenous-origin sediments, movement and diffusion of elements in relation to water flows, and element supplies in connection with thermal and cool upwellings at sea-bottom spreading and subduction areas.

4. *Deep-sea living organisms.* We would like to investigate the various biological processes going on under special physical and chemical conditions of the deep sea. We will give special attention to microorganisms found in the special environment of hydrothermal vents or cold seepages in the deep sea, investigate their physiological and biochemical characteristics, and establish methods of culturing them. Moreover, we will endeavor to understand the characteristics of unique genes and proteins and search for materials useful for man. As stated earlier, we expect the current phase of research and development to extend for the next 15 years. As for large deep-sea organisms, we would like to study taxological and biogeographical characteristics of deep-sea ecosystems in hydrothermal and cold seepage areas, their mechanisms of evolution, and physiological and ecological characteristics.

Table 5. Oceanographic observation and research

- Oceanographic research in the Western Tropical Pacific
- Oceanographic research in the Northern Pacific and Arctic Oceans

In order to achieve the targets mentioned above, we will coordinate operation of SHINKAI 2000, SHINKAI 6500, DOLPHIN 3K, an ROV designed for depths of 11,000 m, and so forth. In addition, for crustal movements and material circulations, we think it important to build networks of long-term observation stations deployed on the ocean floor that would make continuous and wide-area observation possible.

Oceanographic Observation and Research

In our oceanographic observation and research, we aim to understand the role that the ocean plays in global processes of heat transport and materials flux including air-sea interactions, taking the Pacific Ocean and its adjacent seas as our principal field of observation (Table 5). In addition, we would like to develop a numerical simulation method to predict changes in the ocean environment based on sets of data obtained through the studies mentioned above.

During the next decade, we will conduct integrated oceanographic research in the Western Tropical Pacific and the Northern Pacific/Arctic Oceans, two sea areas in which research is urgently called for in relation to the earth's environment. In parallel with this, we would like to establish within the next 10 years an observation network across the Pacific, in a joint effort with the countries concerned. Such a network is considered indispensable to the overall understanding of the processes of the Pacific Ocean in the twenty-first century. Our research in the Western Tropical Pacific will emphasize the mechanism of warm pool generation and dissipation in the Western Pacific, interactions between tropical currents and subtropical gyres including the Kuroshio, and ocean flux in the equatorial upwelling region. The main targets of oceanographic research in the Northern Pacific and Arctic Oceans are evaluation of the effect of the region on the global heat balance and understanding ocean flux in the region.

In order to implement the above research, we must improve numerical simulation models for global ocean circulation and provide observing systems to cover the entire Pacific. The observing systems will combine such modern methods as earth-observing satellites, Lidar, and ocean acoustic tomography systems, as well as existing technologies, such as morring buoy arrays and hydrography by ship. A large-scale research vessel, with capability of observation under harsh weather, will be indispensable to such a program. We will obtain various kinds of satellite data in close cooperation with the National Space Development Agency of Japan (NASDA) and make the information available to researchers after appropriate adaptation for practical use. Lastly, it is our wish to have a research center that will carry out concerted research on the Pacific combining data obtained by satellites and other means.

Table 6. Development and utilization of coastal sea

- Development of highly effective utilization system for coastal sea space
- Development of purification system for coastal sea
- Development of origination system of coastal sea environment
- Development of general management methods for coastal sea
- Countermeasures for local demands

Development and Utilization of Coastal Seas

Coastal seas represent valuable resources for residents. Effective utilization of these areas must be encouraged in response to local needs and in line with the region's long-range perspectives, while paying attention to environment preservation and safety. In Japan, much has been done by varied agencies and organizations concerned. For our part, we would like to set targets as follows (Table 6):

1. We believe it very important to develop new systems that will result in wider, more multipurpose, and more effective utilization of coastal seas including offshore areas, taking into consideration the effect upon the environment. We will develop several new systems, e.g., technology to create calm seas, utilization of ocean energy such as wave energy, large scale aquaculture in the sea and on the sea bottom, as well as underwater access technology and physiology under high-pressure environments necessary to support the above mentioned coastal area utilization systems.
2. Since we think it important to improve the quality of the sea water and the sea bed, we will carry out R&D on purification of coastal seas based on aeration and exchange of seawater.
3. Since we think it will become increasingly important to recover deteriorated environments or create new environments in coastal areas while giving adequate ecological and scenic considerations, we will develop technologies to provide artificial amenity beaches, seaweed zones, tide lands, and coral reefs.
4. Coastal seas will have to be assessed and managed based on scientific data on varied processes occurring there. We will attempt to develop a valid assessment and management system for this.
5. We will strengthen our organizational set-up to carry out advance research jointly with local governments and develop suitable systems to respond to individual local needs quickly and effectively.

Strengthened Environment to Support Research and Development

In order to execute the programs and activities stated earlier, providing adequate support environment is essential (Table 7). Above all, it is of upmost importance to strengthen the human environment. It is necessary to attract capable researchers from around the world and activate research programs by making the institution open to the outside scientific community. For this purpose, we will

Table 7. Basic facilities providing for further studies

- Larger research vessel
- Drilling vessel for deep sea
- Observing system
  - Deep sea monitoring station network system
  - Oceanographic monitoring buoy network system
  - Satellite data high utilization system
  - Supercomputer system
- Pacific Ocean research center

provide needed organizational reforms and new systems. As for the physical environment, I have already mentioned that we will provide a large-size research vessel with a high anti-storm capability to respond to the growing need for oceanic observation under bad weather.

We think it necessary, with a view to promoting research on subbottom strata, to provide a second-generation drilling ship with a riser tube and other capabilities over the existing ship which has been used for the Ocean Drilling Program (ODP). Furthermore, we will develop and deploy an observation network system, consisting of buoys and deep-sea stations, which will permit the monitoring of varied processes occurring from the sea surface to the sea bottom in wide-area, three-dimensioned, simultaneous, and long-term manners. Satellite observation, in the meantime, has been increasing its importance in observing wide areas such as oceans. We will thus provide a facility for effective utilization of satellite data. In addition, we think introduction of a supercomputer is required for ocean dynamics simulations. Lastly, we wish to have what might be called the "Observation and Research Center for the Pacific" which will conduct concerted research on the Pacific Ocean and which is open to the international research community.

Regarding international program participation, we will take part in joint programs such as JGOFS (Joint Global Ocean Flux Study), Inter-RIDGE, and IGBP (International Geosphere-Biosphere Program), and cooperate in establishing GOOS (Global Ocean Observing System). In addition, we may be taking initiatives for some new international programs centering on the Pacific. Furthermore, we would like to promote cooperation with other nations of the Asia-Pacific region to which our country belongs.

For these purposes, we will provide reorganization and new systems as needed.

Closing Remarks

I have spoken about the present status and the future plans of JAMSTEC. For my closing remarks, I would like to touch on what we should do in view of the importance of oceanographic research and development today, and what I am expecting out of this symposium.

You probably feel the same way as I do that everything is deployed on a global scale these days and that the earth is becoming increasingly smaller. It is clear that human beings have only this small planet as a permanent abode, and we have no choice other than to make wise use of its resources while preserving its environment if we are to continue the present comfortable and healthy living standards for ourselves and for posterity. Obtaining comprehensive knowledge of the ocean that occupies such a large portion of the entire earth system is urgently called for all human beings. I am certain that the importance of oceanographic research and development will only keep growing in the years to come.

Implementation of such research and development will require a great deal of qualified manpower and funds. Regrettably, the importance of this research has not been fully understood either by the people at large, or by the government of the country, and as a result, government investment into oceanic research and development has been kept at a low level. Therefore, I think it necessary for those of us concerned with the ocean to rouse national and international public opinion on the importance and potentials of the ocean and to make a strong appeal to the government for increased support for oceanographic research and development. Also, as I mentioned before, organic relations with domestic and foreign institutions are essential to integrated and effective implementation of oceanographic research and development programs. However, such organic relationships cannot be established in a day, and their realization requires a firm will and sustained active efforts.

In this symposium, it is hoped that not only will mutual understanding on the status of each institution and its future plans be enhanced, but vigorous discussions will also be carried out on the targets for future research and on specific ways of collaboration, particularly in areas of deep-sea research and ocean observation where concerted international cooperation is especially important. I will be happy if what we discuss in this symposium leads to further strengthened ties between us and generates certain forms of specific cooperation that will be implemented efficiently in the coming years.

I would like to close my presentation with the hope that this symposium will be the first step toward the realization of organic relationships between the oceanographic research institutions.

Activities of the Ocean Research Institute, University of Tokyo and Its Future Development

Tomio Asai[1]

Key Words. Ocean Research Institute (ORI) — University of Tokyo — Collaboration research — Tansei-Maru — Hakuhô-Maru — Marine science — Oceanography — Fisheries science — Ocean — The Pacific

Summary. The Ocean Research Institute (ORI), University of Tokyo, was founded in 1962 for the promotion of basic and comprehensive research on marine sciences and currently has 16 research divisions in the Nakano campus, Tokyo and a shore laboratory, the Otsuchi Marine Research Center, Iwate Prefecture, covering physical and chemical oceanography, meteorology, geology, geophysics, marine biology, and fisheries sciences. The Institute operates two research vessels, the Tansei-Maru in the seas adjacent to the Japanese Islands and the Hakuho-Maru in the world oceans. These facilities are open to the use of marine scientists throughout Japan.

More than 1,000 researchers in total, including scientists and students from abroad, are involved each year in various aspects of the research efforts of the Institute. The Institute has participated in many international cooperative research projects as a national representative organization. The ORI is also engaged in the academic program of the graduate school of the University of Tokyo. As of 1991, 30 out of the 80 graduate students of the Institute come from Asia, Europe, and the United States.

The ORI is developing towards being a core institution for oceanic research and education in the world as well in Japan. Establishment of an international center for ocean sciences is one of the important future plans of the ORI to advance further international cooperative research now being carried out with many countries, especially those in the Western Pacific region.

Recently research activities of the ORI are developing toward a new direction, in addition to individual traditional research fields. Activities include a study of physical and biogeochemical processes resulting in circulation of energy and

[1] Ocean Research Institute, University of Tokyo, 15-1, 1-Chome, Minamidai, Nakano-ku, Tokyo, 164 Japan

material in the ocean as part of the interacting components composing the planet Earth.

Introduction

The Ocean Research Institute (ORI), University of Tokyo, was founded in 1962 for the promotion of basic and comprehensive research on marine sciences. At that time Professor Koji Hidaka served as the first director. The physical oceanography and submarine sedimentation divisions were the two first research divisions with which the Institute started. As the years passed, other research divisions were added. Currently it has 16 research divisions in the Nakano campus, Tokyo and a shore laboratory, the Otsuchi Marine Research Center, Iwate Prefecture, covering physical and chemical oceanography, meteorology, geology, geophysics, marine biology, and fisheries sciences. The Institute operates two research vessels: one is the Tansei-Maru in the seas adjacent to the Japanese Islands and the other is the Hakuhô-Maru in the world oceans. These facilities are open to the use of marine scientists throughout Japan.

More than 8,000 researchers in total person-days, including scientists and students from abroad, are involved each year in various aspects of the research efforts of the Institute. The Institute has participated in many international cooperative research projects as a national representative organization. The ORI is also engaged in the academic program of the graduate school of the University of Tokyo. In 1991, 30 out of the 80 graduate students of the Institute come from Asia, the United States, and Europe.

The ORI is developing towards being one of the core institutions for oceanic research and education in the world. Establishment of an international center for ocean sciences is one of many important future plans of the ORI to advance further international cooperative research now being carried out with many countries, especially those in the Western Pacific region.

Recently research activities of the ORI are developing towards a new direction, in addition to individual disciplinary research fields. Activities include a study of the physical and biogeochemical processes that result in circulation of energy and substances in the ocean as part of the interacting components composing the planet Earth.

Organizational Structure of the Institute

The organizational structure of the ORI, as shown in Fig. 1, consists of 16 research divisions, research facilities, and an administration office. Here a research division consists of one professor, one associate professor, two research associates, and two technical staff members on average. In 1990, the Institute had approximately 200 personnel consisting of 65 scientists (professors, associate professors, and research associates), 33 administrative officials, and 100 technical staff

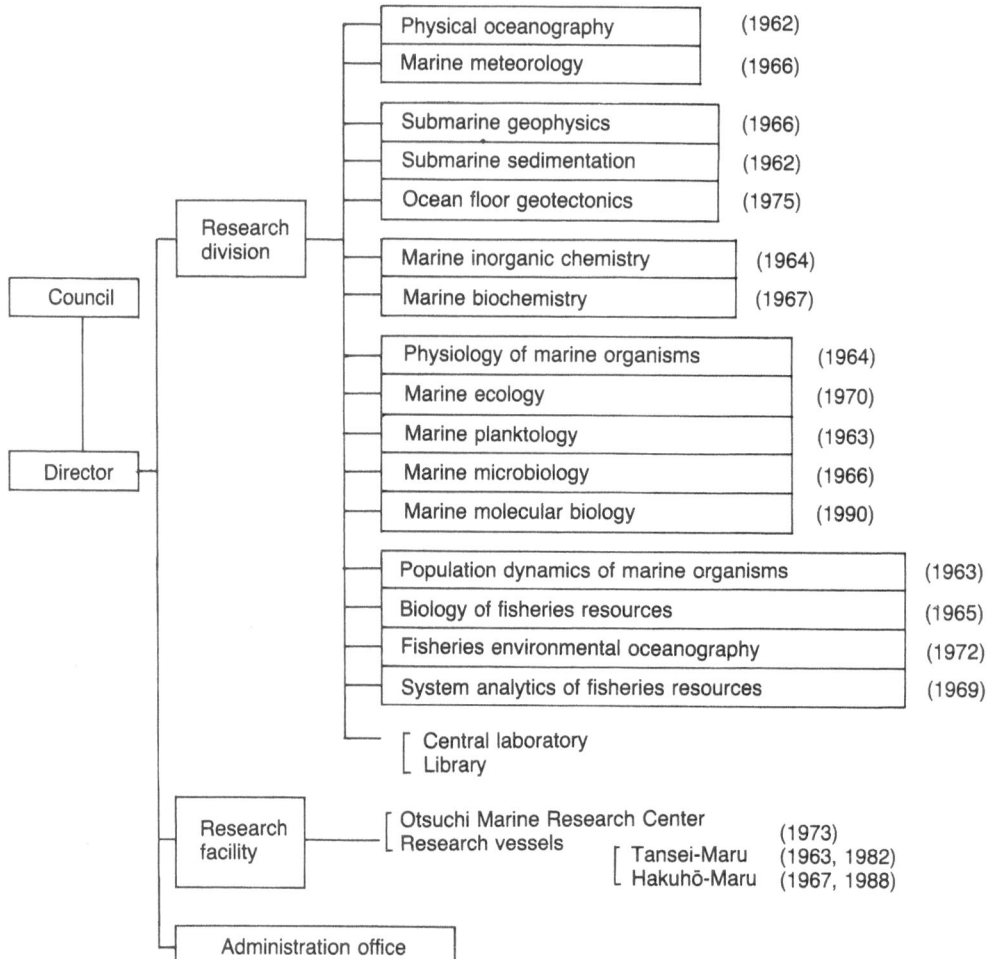

Fig. 1. Organization of the institute

members including crew members (Table 1). Graduate students including foreign students numbered 80–90. The annual budget was ¥2,800 million (including ¥1,260 million for personnel and excluding ship construction). All necessary expenses are supported mainly by the Ministry of Education, Science and Culture, Japan.

The Institute was designated as a collaborative research institute, and conducts both its own research and collaborative research with other individuals and institutions in Japan and elsewhere. The collaborative research consists of (1) joint research between scientists in universities and national institutes in Japan and staff members of the ORI, (2) organization of scientific meetings on current

Table 1. Personnel as of August 1990

	Professor	Associate professor	Research associate	Administrative official	Technical staff	Others	Total
Research divisions	16	16	28		25		85
Research Tansei-Maru					23		23
vessels Hakuhô-Maru				1	43	(1)	44
Otsuchi Marine Research Center	1		4	2	1		8
Research facilities					6		6
Administration office				30	2		32
Total	17	16	32	33	100	(1)	198

( ) Concurrent post

topics in ocean science, and (3) use of research facilities (research vessels and Otsuchi Marine Research Center).

Two advisory bodies, the council and the steering committee, were established to conduct the collaboration research smoothly. These committees are made up of persons with knowledge and experience from outside the Institute as well as of staff members from within the Institute.

Research Facilities

Central Laboratories

The central laboratories have been set up to provide large equipment for common use by researchers in the Institute, including (1) a radioisotope room, (2) computing facilities, (3) an aquarium room, (4) a laboratory for cultivation of microalgae and bacteria, (5) a P2 laboratory for molecular cloning, (6) a sample and specimen room, (7) low-temperature rooms, (8) a photographic room, (9) a machine shop, and (10) instrument rooms (scanning electron microscope, ultracentrifuge, mass spectrometer, chromatograph, X-ray analysis apparatus, and so on).

Research Vessels

The ORI has two research vessels, the Tansei-Maru (Fig. 2) and the Hakuhô-Maru (Fig. 3) for multidisciplinary research activities in the oceans. The new Tansei-Maru was put into service on October 15, 1982 to replace the old Tansei-Maru. It is engaged in 18–20 scientific cruises every year in the seas adjacent to the Japanese Islands. The old Tansei-Maru, completed on June 20, 1963 with a gross tonnage of 257.69 tons, made 399 scientific cruises with 414,198 km of total cruising distance. A total of 3,783 scientists (1,652 from inside the Institute and 2,131 from outside organizations) participated in these cruises on board this vessel during its 20 years' working period of 1963–1982.

Fig. 2. Tansei-Maru

Fig. 3. Hakuhô-Maru

The new Hakuhô-Maru was launched on October 28, 1988 and put into service on May 1, 1989. Ten laboratories including multi-purpose dry, semi-dry and wet laboratories, a freezing laboratory, and a clean room encompass 387.71 m^2, accomodating a Seabeam, NOAA-GMS-BS satellite receiver, isotope treatment facilities, acoustic biomass survey system, conductivity-termperature-depth profiler (CTD), Doppler current meter, liquid N2 tank, multichannel seismic profiler, 3.5-kHz subbottom profiler, acoustic transponder system, ocean-floor imaging system, gravimeter, high-precision gyrocompass, and data processing system linked with navigation instruments by LAN and Ethernet. A research room equipped with word processors and meeting facilities is also installed. It was built to replace the old Hakuhô-Maru that had a size of 3,200 gross tons, was completed in March 1967, and operated until the end of the fiscal year 1988. Between 1967 and 1988 the old Hakuhô-Maru completed 99 scientific cruises (together with 9 sea trial cruises) with a total cruising distance of 1,178,035.4 km. A total of 3,643 scientists (1,332 from inside the Institute and 2,311 from outside organizations) utilized this vessel during its 22 years' working period of 1967–1988.

Both the Hakuhô-Maru and Tansei-Maru are open to all scientific personnel as are the other research facilities in the Institute. About 60% of the shipboard participants have come from other organizations out of the Ocean Research Institute. The two vessels have participated in a number of international cooperative programs such as CSK, WESTPAC, IBP, GDP, GARP, BIOMASS, IPOD, ODP, KAIKO, WCRP, IGBP and so on.

Otsuchi Marine Research Center

The Otsuchi Marine Research Center was established in 1973 at Akahama, Otsuchi, Iwate Prefecture. This Center has been open to the use of visiting scientists since 1979.

The Center is located on the northern shore of Otsuchi Bay in the northeastern part of Honshu, Japan (Fig. 4). The Tsugaru Warm Current flows along the coast from the north, and the Oyashio and Kuroshio Currents meet off the coast. The water of Otsuchi Bay is formed by the mixing of offshore water and water from rivers flowing into the inner part of this bay. Such a marine environment provides a unique opportunity to study physical and biological phenomena in both cold and warm water masses.

The Center consists of a three-storey main building with an area of 2,194 m^2, a two-storey dormitory of 330 m^2 for 20 visiting scientists, a warehouse of 390 m^2, and 30 outdoor culture tanks. The main building houses five laboratories for research in physical, chemical, geological and biological oceanography, and fisheries science, a radioisotope laboratory, an experimental aquarium room, culture rooms regulated to temperatures of 5°, 15° and 25°C, five constant temperature rooms, five research rooms for scientists, a library, a meeting room, a workshop, and an administration office.

Two hundred tons of filtered sea water are supplied to the tanks and laboratories each hour. Three lines of running sea water of up to 2 tons/h with temperature controlled in the range of 8°–27°C are available in the experimental aquarium room in all seasons.

The Center is equipped with a variety of specialized apparatus including current meters, a long wave recorder, sea water temperature recorders, a STD salinometer, spincotype ultracentrifuges, electron and scanning electron microscopes, a data analyzer, and an underwater TV and VTR. The Center has three research boats, the Yayoi (16 tons) with a winch and echo sounders, the Rias (2.1 tons), and the Challenger (1.0 ton).

Fig. 4. Location of the center and institute

Current and Future Research Activities

The basic purpose of the Institute is to conduct research directed towards a better understanding of the complex problems of marine science. Until April 1975 the following 15 research divisions and the Otsuchi Marine Research Center had been established. A new research division on marine molecular biology has just started in 1990.

Physical Oceanography

The dynamics of the following subjects are investigated on the basis of direct measurement from research vessels, analysis of historical data, and theoretical and numerical modeling. An ultimate goal of the research is to understand the physics of ocean variations and to predict them on the basis of a dynamic model.

1. The Kuroshio variation
2. Deep circulation in the western North Pacific Ocean
3. Bottom currents in deep trenches
4. Meso-scale eddies and oceanic waves
5. Subtropical ocean and the El Niño
6. Ocean mixed-layers

Marine Meteorology

The atmosphere and the ocean interact with each other in various ways and on various time and space scales. A primary objective in our studies is to understand the mechanisms of these interactions. In particular, dynamic features of the following problems are under investigation theoretically, and experimentally, as well as by the analysis of observational data.

1. Airmass transformation processes over the ocean and the heat balance
2. Meso-scale atmospheric disturbances associated with heavy precipitation
3. Extratropical cyclones, blocking, and the baiu-front
4. Local wind systems in coastal regions

Submarine Geophysics

The following research is now in progress in order to elucidate crustal and subcrustal structures under the oceans and the physical processes of plate motions.

1. Marine geodetic and geophysical studies based on gravity measurements at sea
2. Marine geodetic studies based on satellite altimetry
3. Electromagnetic explorations on the seafloor
4. Long-term and precise geophysical observations at ocean bottom stations
5. Superconducting gravimetry
6. Submarine geophysics in the Antarctic

Submarine Sedimentation

Marine sediments are the products of complex interactions between tectonics, paleoenvironments, and sea level changes. The major research goal is to investigate processes of marine sedimentation, origins of sedimentary sequences, and crustal evolution. Thematic objectives of interest include:

1. Physical sedimentology
2. Origin of sedimentary sequences
3. Crustal evolution
4. Global environmental events
5. Participation in international cooperative programs of research

Ocean Floor Geotectonics

The structure and dynamics of the ocean floor are studied in this division. The principal objectives are to investigate the dynamics of the oceanic lithosphere and its driving forces on the basis of the integration of the geophysical and geological data.

1. Measurements of the marine geomagnetic anomalies, improvement of geomagnetic reversal time scales and studies of the global history of sea floor spreading aiming at its inversion to mantle dynamics
2. Evaluation of the structure of the earth's interior, based on data on topography, sediment stratigraphy, seismic velocities, gravity anomalies and heat flow
3. Studies of the rheological properties of mantle rocks based on experimentally observed deformation, flow, and the microscopic texture of samples
4. Measurement of the remnant magnetization of piston cores and ODP (Ocean Drilling Program) samples, and determination of geomagnetic field variations, ocean floor spreading and sedimentation rates, and studies of volcanic rocks and hydrothermal minerals to clarify the history of volcanic activity and paleoenvironments on the ocean floor
5. Petrological and paleomagnetic studies of samples dredged and drilled from the oceanic basement, seamounts, and ocean islands
6. Studies of the evolutionary history of island arcs, based on the paleomagnetic properties of ophiolites
7. Studies of the deep sea floor using deep sea cameras and submersibles

Marine Inorganic Chemistry

The major goals are to clarify the distribution and circulation of various components within the oceans and through the lithosphere-ocean-atmosphere system and to draw together a picture of the geochemical evolution of the oceans. In this context, evaluation of the capacity of the oceans as a sink for man-made pollution is also a concern of this division. In order to achieve these goals, studies are currently in progress on the distribution and circulation of inorganic components, including stable and radioactive isotopes in seawater, bottom sediments, and the underlying oceanic crust. Special emphasis is placed on the following projects:

1. Submarine volcanism and hydrothermal activity
2. Distribution and circulation of various components and isotopes in the oceans
3. Development of analytical and observational techniques

Marine Biochemistry

The distribution and circulation of carbon, nitrogen, phosphorus, sulfur, and silicon in the sea are regulated by the activities of various kinds of organisms. These elements are present in volatile, dissolved, or suspended forms, and thus their biogeochemical cycles in the sea are closely associated with those in the atmosphere and lithosphere. Because of its large capacity, the sea plays a crucial role in maintaining the global cycles of these elements. Research in this division is concerned primarily with the dynamics of these biophilic elements in marine environments. Emphasis is placed on the identification of the various types of biochemical processes operating in seawater and submarine sediments and their quantification. The following research projects are currently in progress:

1. Cycles of nitrogen, carbon, and phosphorus cycles and their interactions
2. Isotope fractionation and its application to the analysis of biogeochemical processes
3. Biochemical and physiological studies of nitrogen and carbon metabolism

Physiology of Marine Organisms

Scientists focus their attention on the resolution of adaptive mechanisms of many and diverse marine life forms to various environments. Current studies have paid special attention to the relations among fish migration and osmoregulation. Several scientists are working on the adaptation of marine plankton to environmental conditions of varying light intensity and nutrient availability.

1. Fish migration and osmoregulation
2. Adaptive mechanisms of phytoplankton to marine environments

Marine Ecology

Ecology of marine organisms in general, especially benthic animals in various bottom ecosystems from intertidal to hadal (trench) depths and from tropical waters to the polar seas. Benthic communities and specific populations are described and their dynamics are investigated using research vessels and various kinds of monitoring or sampling gears. Current main projects are:

1. Estimation of the standing crop of benthos and evaluation of factors related to the abundance and structure of benthic communities
2. Reproductive and physiological ecology of benthic organisms
3. Role of benthos in the bottom community
4. Hydrothermal and cold-seep ecosystems
5. Biogeographical and evolutionary studies of benthic organisms

Marine Planktology

This division is concerned with ecological and physiological studies, in areas such as the distribution, production, and temporal variations in density of plankton and micronekton in the sea. The role of plankton in the function and structure of the marine ecosystem as a whole is also studied. The following research projects are currently in progress:

1. Biological production in the ocean
2. Biology of key species of plankton
3. Blooms, mass propagation of plankton, and red tide phenomena

4. Experimental studies on plankton and micronekton
5. Biochemical studies on plankton

Marine Microbiology

Research is being carried out to clarify the physiological and ecological characteristics of marine bacteria and other microorganisms and their roles in the process of interaction between microorganisms and marine animals and plants. The current research projects are as follows:

1. Ecological studies of marine microorganisms
2. Microbiological and chemical studies on the process of decomposition of organic matter in the sea

Marine Molecular Biology

Adaptation of marine organisms to environmental changes is one of the most fundamental biological phenomena. Clarification of its mechanisms will be crucial for an understanding of marine organisms. Our major interest focuses on this problem using molecular techniques.

1. Molecular mechanisms of fish migration
2. Association of the neurosecretory system in adaptation
3. Adaptational roles of symbiosis in marine animals

Population Dynamics of Marine Organisms

This division was originally concerned with the population dynamics of marine organisms to establish a basis for the scientific management of fishery stock. Biochemical or molecular genetic analyses were recently introduced to understand populations and species in marine organisms in terms of population genetics.

1. Genetic studies of populations in commercially important species
2. Genetic distance and molecular phylogeny of marine organisms
3. Analyses of fisheries data
4. Theoretical studies on the dynamics of fish stocks

Biology of Fisheries Resources

The fundamental ecological and physiological factors underlying both the fluctuation of naturally occurring populations and of the artificial propagation of fisheries resources are being examined in order to establish ecologically sound principles of resource management. The following research projects are currently in progress:

1. Life histories of fishes
2. Migration mechanisms of salmonoid fishes and eels

3. Studies on the mass mortality of larval fish through mark-recapture techniques
4. Metamorphosis of leptocephali of the family Anguillidae
5. Behavioral development in larval fishes
6. Studies on the life history strategies of brackish and coastal water fishes
7. Studies on fish otoliths

Fisheries' Environmental Oceanography

Structures and their variations in oceanographic environments are investigated to establish a basis of the forecasting and the management of fisheries. The research activities directed toward these objective are as follows:

1. Investigations are being carried out on oceanographic structures related to the distribution and movements of fish schools in fishing grounds based on data from direct current measurement, satellite thermal imagery, and other hydrographic surveys.
2. Year-to-year variations in oceanographic conditions around spawning and nursery grounds related to fluctuations in the abundance of marine biological resources are being investigated. The processes of transport and dispersion of fish eggs and larvae are also being studied, using different types of drifters.
3. Numerical and hydraulic model experiments on meso-scale eddies, frontal disturbances, and mixing processes are being conducted. Numerical modeling of primary production and fisheries' ecosystems are also being studied.
4. Studies are being done on the coastal environment, including descriptive and dynamic studies on tide-induced currents, density currents, and wind-driven currents in estuary, bay and shelf regions. The role of these currents in water exchange processes and dispersion of substances are the main areas of investigation.

System Analytics of Fisheries Resources

Eventual objectives are to develop techniques to estimate the abundance of fishes and micronekton and to observe ecological features through the application of image processing and artificial intelligence techniques. Also research to elucidate the mechanisms influencing the stock size of these organisms is being carried out. The research activities directed toward these objectives are as follows:

1. Applications of hydro-acoustic techniques to estimations of fish abundance and behavioral studies
2. Development of an underwater observation system
3. Analysis of schooling behavior of fish
4. Studies on the efficiency of equipment for quantitative sampling of marine animals
5. Studies on the applications of image processing and artificial intelligence techniques
6. Analysis of fishing conditions using satellite information

Table 2. International cooperative studies

World Climate Research Programme (WCRP)	1987–1990
Ocean Drilling Program (ODP)	1985–1994
French-Japanese Cooperative Scientific Studies on Japanese Trenches (KAIKO Project)	1984–1986
International Lithosphere Program (ILP)	1985–1990
Biological Investigation of Marine Antarctic System and Stocks (BIOMASS)	1977–1986
Species identification of plankton by pattern recognition	1983–1985

Otsuchi Marine Research Center

The investigations carried out by staff members and visiting scientists in this research center embrace a wide variety of fields in marine science, including marine ecology, physiology, taxonomy, physical oceanography, submarine geophysics and sedimentation, marine chemistry, biology, and fisheries science. The number of visiting scientists is about 1,500 person-days per year. Investigations being conducted mainly by the present staff members are as follows:

1. Ecosystem study of Otsuchi Bay
2. Life history strategies of the key fish species in Otsuchi Bay with special reference to restocking mechanisms during the early life history
3. Studies on the community structure and life history of epifaunal crustaceans in the Sargassum Zone, with special reference to the importance of the Sargassum Zone for the restocking process of commercially important fish species
4. Physiological and ecological study of seagrasses through field and culture experiments under various environmental conditions
5. Ecological studies of mesopelagic fishes off the Tohoku region
6. Taxonomical, ecological, and physiological properties of the aerobic photosynthetic bacterium *Erythrobactor*, and the evolutional origin of the bacterium
7. Mechanisms of oceanic and atmospheric variabilities in areas off the Sanriku Coast, northern Honshu

International Cooperation

The ORI has participated in several national and international cooperative projects since its establishment. These projects also include interdisciplinary projects as well as those belonging to specific disciplines. The Institute serves as a coordinating center for the participation of scientists from all over Japan.

International Cooperative Studies

International cooperative studies sponsored by the Ministry of Education, Science and Culture after the 1985 fiscal year are summarized in Table 2. Besides these activities, geochemical research on hydrothermal plumes near Roihi island in Hawaii (on board the R/V Hakuhô-Maru of ORI in 1985) and biological research near the hydrothermal vent at the spreading center of the Mariana back arc basin (on board the Atlantis II and Alvin of SIO in 1987) were carried out. The Regional Program of the Western Pacific (WESTPAC) was organized under the Intergovernmental Oceanographic Commission (IOC/UNESCO) following the Cooperative Study of the Kuroshio (CSK) which finished in 1980. WESTPAC was a program group which made up research plans over the Western Pacific Ocean, and conferences were held three times until 1987. ORI has been making efforts to maintain WESTPAC activities. In particular, from 1981, programs of cooperative research and education have been carried out on board the R/V Hakuho-Maru and participants included researchers from countries included in WESTPAC (which is now the IOC Subcommission).

Scientific Cooperation Programs with Southeast Asian Countries

Since 1978 the Japan Society for Promotion of Science (JSPS) has been carrying out special bilateral exchange programs with counterpart institutions in Southeast Asian countries, with the intention of developing scientific cooperation between Japan and these countries. One unique characteristic of the bilateral programs is that they have adopted what is called a "core university system," in which designated universities play a leading role in conducting cooperative activities in selected academic fields. In the core university system, a cooperative network is established in each academic field through negotiations between JSPS and its counterpart institutions. These networks are composed of a core university and cooperating universities and scientists on each side.

The Ocean Research Institute, as a core university sponsored by JSPS, plays a role in coordinating cooperative activities in marine science with the cooperating universities and scientists in Japan, and also in jointly planning and implementing these programs with counterpart core universities in the other countries (currently Indonesia, Thailand, and Malaysia).

Grants-in-Aid for Overseas Scientific Research

The following studies were made over the past few years through Grant-in-Aids sponsored by the Ministry of Education, Science and Culture, Japan:

1. Studies on the mechanisms of marine productivity in the shallow waters around the South China Sea
2. Studies on the dynamics of the biological community in tropical seagrass Ecosystems
3. Surveys of oceanographic and fisheries studies on coastal exploitation and management in France
4. Investigation of fisheries and oceanographic research in Australia
5. Research and investigation into physical, chemical, and geological characteristics of ocean drilling holes
6. Osmoregulatory mechanisms of the crab-eating frog
7. A study on developmental process and material flow in tropical seagrass ecosystems

Table 3. Students

	Graduate student				Research students	JSPS postdoctoral research fellow
	School of Science		School of Agriculture			
	Ms.	Dr.	Ms.	Dr.		
1987	11	20	6	19	27	2
	(2)	(4)	(3)	(9)	(7)	
1988	15	24	4	17	21	2
	(4)	(9)	(1)	(6)	(7)	
1989	12	29	10	17	14	5
	(3)	(11)	(1)	(7)	(3)	
1990	16	31	11	11	13	4
	(4)	(11)	(1)	(3)	(4)	

( ) Foreign students

Academic Exchange Agreement

Academic exchange agreement with Maryland University, Woods Hole Oceanographic Institution, Scripps Institute of Oceanography, School of Ocean and Earth Science and Technology, and the University of Hawaii have already been arranged.

Educational Activities

Almost all of the professors and associate professors of the Institute are members of the Graduate School of either the Faculty of Science (geophysics,, chemistry, zoology, botany, geology, geography) or the Faculty of Agricultural Sciences (fisheries), the University of Tokyo, to which they are affiliated. They are engaged in an academic program through course work and guidance of graduate students. Special lectures in oceanography are also given to undergraduate students in the College of Arts and Sciences during the second (winter) term. In addition, the institute accepts domestic and foreign research students and research fellows.

Eighty graduate students including thirty foreign students from Asia, Europe, and the United States were registered in April, 1991 (Table 3).

Publications

The ORI issues the "Bulletin of the Ocean Research Institute, University of Tokyo," "Otsuchi Marine Research Center Report," and "Preliminary Report of the Haku-Maru Cruise." In addition an annual list of titles of scientific articles published by the staff and students is issued as the "Publication List" and the number of articles registered as of December 1990 had reached 3,272.

USA

Woods Hole Oceanographic Institution: Status and Plans

CRAIG E. DORMAN[1]

Key words. Oceanography — WHOI — Marine science — Environment — Global change — Ships

Summary. Woods Hole Oceanographic Institution (WHOI) was founded in 1930 as a private, nonprofit research laboratory dedicated to prosecute oceanography in all its branches. The impetus for its founding was a study by the United States National Academy of Science which pointed out the importance to the nation of the then fledgling field, and recommended the establishment on the east coast of an institution which could both develop its own critical mass of multidisciplinary scientific talent, and make the heavy investment in ships and facilities needed to spur on the field as a whole. Its special expertise was to be its ability to make and interpret state-of-the-art measurements at sea.

WHOI today remains true to its initial charge, both to conduct ocean science and education at the frontier of knowledge and to support the oceanographic community as a whole. WHOI in 1991 has a staff of approximately 1,000, including 140 on the scientific staff and 130 students in the graduate joint program with MIT, and an annual budget of over $70 million. We operate three major research ships, two national facilities — the 4,000-m-capable manned submersible ALVIN and the new Ocean Science Accelerator Mass Spectrometer — and a wide range of seagoing vehicles and instrumentation.

WIIOI's forte is the study of ocean process based on measurements at sea. Five disciplinary departments — Biology, Chemistry, Geology and Geophysics, Physical Oceanography, and Applied Ocean Physics and Engineering — are complemented by centers devoted to coastal processes, marine policy, and deep ocean exploration. One of WHOI's special skills is the development of instrumentation for at sea data collection, and many of today's standards originated in our laboratories.

[1] Woods Hole Oceanographic Institution, Woods Hole, MA 025473, USA

WHOI's future plans are motivated by the intersection of the interests of its scientists and society's concerns with the global environment. Our basic institutional thrust will be to increase our ability to observe and understand the oceans throughout their volume and continuously in time; thus we need to extend our ability to operate in high sea states, in the Arctic, and at great depths, and to develop ways to bring ashore data collected by unmanned platforms which remain continually at sea. We have new initiatives in air-sea interaction to help improve our understanding of coupled atmosphere ocean dynamics and exchange, and in instrumentation to develop the next generation of state-of-the-art ocean measuring systems.

As an institution whose interests are global, WHOI values its collaborations with scientists from many nations. We are particularly proud of our relationships and joint programs with colleagues at JAMSTEC, both in the development of new instrumentation and in seagoing science including the joint initiative on Arctic research.

Woods Hole Oceanographic Institution (WHOI) was founded in 1930 as a private, nonprofit research laboratory dedicated to prosecuting oceanography in all its branches. The impetus for its founding was a study by the United States National Academy of Science, which pointed out the importance to the nation of the then fledgling field and recommended the establishment on the east coast of an institution which could both develop its own critical mass of multidisciplinary scientific talent and make the heavy investment in ships and facilities needed to spur on the field as a whole. Its special expertise was to be its ability to make and interpret state-of-the art measurements at sea.

The village of Woods Hole in the town of Falmouth on the southwest end of Cape Cod was selected as the location for this new institution because of its deep water harbor, proximity to the open waters of the North Atlantic, ready access to the intellectual centers of Boston and Cambridge, and presence in the village of a companion institute, the Marine Biological Laboratory (MBL). In the United States prior to World War II, support of research and higher education primarily was the responsibility of industry and philanthropists. Thus, in spite of its national charter, WHOI was established as a private corporation. An initial $3 million grant from Rockefeller Foundation enabled our first director, the Harvard marine biologist Henry Bryant Bigelow, to erect a waterfront building and dock on land bought from MBL, to commission a 44-m, 460-ton double-ended steel ketch he christened ATLANTIS, and to support several years of seasonal expeditions in the Gulf Stream, Gulf of Maine, and other Atlantic waters.

In his 1950 retrospective report to the Trustees summarizing his own 10 years as our second Director, Columbus O'Donnell Iselin noted that when he relieved Dr. Bigelow in 1940

we were a laboratory for individual investigators...Although our program was well integrated and we made good use of our facilities, the sorts of problems that could be attacked effectively were rather limited. We had no machine shop. Except for an old fashioned radio on ATLANTIS, we ignored all electronics equipment. Only one member of our staff, Dr. Watson, had an advanced degree in Physics. None was skilled in mathematics. There was little chance of fitting people together into research teams so that a number of skills and techniques could be brought to bear on a single problem. Our science was largely descriptive, rather than experimental.

Iselin went on to describe the dramatic changes in the field which occurred during the 1940s:

The war brought physics to oceanography and it showed us the great advantage of research teams. We have never really organized teams as an industrial research laboratory does. They have just grown and people are free to drop out from one team and to join another. The research is not directed nor are the problems specified. All we do is to make it easy for the teams to form. So far as possible each qualified investigator has some project of his own, but also on a part time basis he contributes what he can to other studies. Of course, in every laboratory there are individuals who like to work alone and this we do not discourage. We simply hope that such people will come to find the benefits of working with others . . . Furthermore, in the case of oceanography, the minimum effective size of the laboratory is also very much conditioned by the field work. People do not want to be at sea continuously, nor is it good to have one group carrying out the observational program and another working up the results.

In summarizing his tenure, Iselin quoted from a 1945 Presidential Award for accomplishments: ". . . 'which saved a large number of our ships . . .' If we are called on again, it will not be easy to do better"; "Our policy at the outset (of the war) was to place our resources at the disposal of the government"; "A number of the key men at . . . government laboratories first went through the mill here at Woods Hole. Thus, not only did we initiate much of the present large research program in the marine sciences, but we also helped train a number of the more productive investigators."; "The increased scale of operation has made it possible since the war for the Institution to help in the establishment of other oceanographic groups," notably what evolved into Columbia's Lamont-Doherty Geological Observatory's ocean-related program, started by Maurice Ewing from WHOI.

In these early years, even in spite of the pressures of the war, the Institution remained very loosely organized. Notably, the Director and staff had extensive latitude in expenditure of Government funds. The money came to the Institution to pursue a basic program of science and to support the process of investigation. The sponsors had faith that the scientists themselves were best at selecting and following paths of inquiry, along lines of very loosely described needs, such as acoustics and anti-fouling.

Both the direct and indirect contributions of scientists to the war effort convinced our government of the value of science to national security. To ensure continuity of this resource, at the end of the war the Federal Government took upon itself the responsibility for funding the process of science, and estalished the National Science Foundation (NSF) to act as sponsor and coordinator. Philanthropy and industry came to play a less dominant role in maintaining the intellectual strength of the nation.

The government decisions to fund science, and the use by NSF of a peer review process to judge quality proposals, created fertile ground for the development of an extremely strong United States science base. For the past 40-plus years, it has provided a sound foundation for stimulation of individual creativity. Over time this process has become highly structured, but it remains an effective way of providing funding directly to individual investigators, often within the context of large multi-investigator initiatives developed by the scientific community in consort with the sponsors.

As our scientific practice and its funding processes evolved, the nature of our field changed and the number of its practitioners grew. When Iselin highlighted accomplishments of the leading scientists at WHOI in 1950, he mentioned 9 "misters" and only 5 "doctors." Most of the staff came to oceanography from other fields, and few were formally educated in it. Today, all 140 of our scientific staff members (except Hank Stommel, our most senior and renowned member) have earned PhDs, as have most of the 150 of our technical staff who work in their laboratories. The vast majority of them received their graduate training in oceanography at one of our sister institutions in the United States or abroad, or in our own Joint Program with Massachusetts Institute of Technology.

Not only did WHOI quadruple in size between 1950 and 1980 — we have been relatively steady at a staff plus student level of about 1,000 since then — but the practice of our science spread during those years to many institutions. Today, we have nearly 60 institution members in the University National Oceanographic Laboratory System, an organization of users of oceanographic ships, and in the Council on Ocean Affairs, a group newly formed to help educate our federal legislators. If one drops down a size-tier from those practicing science in the world oceans to marine labs concentrating on local coastal, riverine or lacustrine problems, one finds literally hundreds of participating organizations.

While similar growth and maturation has occurred in many fields of science in the United States oceanography remains unusual, if not unique, in two respects. First, federal support for our field grew so rapidly in the 1970s during the International Decade of Ocean Exploration that it outpaced the expansion of teaching positions within our research universities. Thus, many of our scientists are supported by what we call soft money: their positions are unendowed or unsupported by guaranteed funding from teaching. Particularly at WHOI, but also at most other major oceanography institutes, these research practitioners must fund themselves as well as their laboratories and field work by competitive federal grants. Second, our national expertise in ocean science rests firmly in academia. While some federal agencies maintain ocean laboratories or centers, there is no "National Ocean Science Center," and even the infrastructural support for the science — the ships and primary instrumentation — is operated by academic research institutions and universities.

WHOI today plays a major role in this national structure, remaining true to its initial charge, both to conduct ocean science and education at the frontiers of our knowledge, and to support the oceanographic community as a whole. We have of course changed over time in response to changes on the national and world scene. With NSF funding of individuals, for example, the Director no longer "directs." Sheer size dictated even in the 1960s that we formalize our organization beyond that of the happily unstructured days of Bigelow and Iselin, and today, we have departments of Biology, Chemistry, Geology and Geophysics, Physical Oceanography, and Applied Ocean Physics and Engineering. These departments promote intellectual interaction among scientists of common disciplinary interest and resource needs. We also strive through both physical and operational procedures to maximize interdepartmental interaction and eliminate administrative barriers to collaboration. In addition, WHOI has an interdisciplinary center to support coastal studies; a center for the study of marine policy, where economists, lawyers and statisticians interact with the rest of our scientific staff on issues related to man's interaction with the oceans; and a center for marine exploration which promotes the scientific use of our deep ocean remotely operated vehicles.

To complete this thumbnail description, our physical plant has grown from Bigelow's building and dock to comprise some 50 structures on over 200 acres. We operate three major ships for ocean science: OCEANUS, KNORR — which recently completed a major modernization — and ATLANTIS II, mother ship for Alvin, our 4,000-m manned deep submersible now in its 28th year of operation. We also have two coastal craft and a variety of remotely operated vehicles, and an exceptional set of shops, facilities, and people for operating and maintaining these sea going facilities in support of science. We operate the new NSF-supported National Ocean Accelerator Mass Spectrometery as well as a variety of specialized facilities — from an ion microprobe to a rigging shop to a protein and nucleic acid chemistry center — which are accessible to scientists from around the world. One of our special skills is the development of instrumentation for at-sea data collection, and many of today's standards — for example the CTD, SOFAR floats, current meters, and sediment traps — originated in our laboratories.

Finally, although I have described the context for our Institution in national terms because that is our organizational heritage and principal funding source, we are truly international in composition and interest. We historically have been a "blue-water" global ocean institution; we house the JGOFS and RIDGE planning offices, and the WOCE Program Hydrographic Office; our staff and students hail from over 35 nations; we routinely entertain visiting investigators from virtually all maritime nations with ocean science programs; and we are active in promoting and supporting international collaborative science in all the world's oceans and seas.

I have dealt at some length with our heritage because I believe it is particularly important to reflect on rationale and historic forces leading to the current status when major changes are expected. Certainly this is a period of much change on both global and national levels, and I expect all the speakers at this conference will address this point. Indeed, having devoted my entire adult life to national security as I understood it as a naval officer, when I prematurely relinquished my flag to become WHOI Director in 1989 I did so with a conviction that our fundamental concepts of security as a global population in nation states were undergoing drastic evolution, and that the oceans would play a major role in this process and in our future views of our health, security, and welfare.

Simply stated, problems long apparent but perceived as secondary in terms of their threat to our survival as nations suddenly have become even more important — on both regional and global scales — than superpower confrontations. While military and political interactions remain very significant international forces, issues of international economic competitiveness, education and its long term impact on our ethos and industrial viability, and the effect we have had and continue to have on the environment are becoming increasingly central to our identity, strength, and health as nations. Even in those geographic areas undergoing most fundamental and rapid political change, environmental concerns will play a major role in decisions on economic and administrative restructuring. We have started to use the term "environmental security" to describe these concerns, in the sense of the impact of one nation's activities on others' environmental characteristics. We have begun the study and discussions needed to give sufficient form to these concerns that we can discuss them with. commonly understood terminology and rationally debate approaches to resolving them on the international level.

In the United States, these global trends have combined with our own unique concerns to create an environment for science that has forced us to seriously examine our priorities as an institution. Let me briefly enumerate some of the United States national issues as I see them. First, I believe that the national consensus that led at the end of World War II to the commitment to federal support for basic science as a major national priority is eroding. The very global environmental concerns that to us, here, imply a need for enhanced emphasis on fundamental studies, are seen by much of the populace as indicative of a failure of science to avoid the problem. I have heard science referred to as a "bottomless black hole" in our congress. Our administration leaders, while praising science and supporting NSF budgetary increase, mistakenly point to engineering and system developments like satellites, as examples of scientific accomplishment. The cost and complexity of large programs like the human genome project and the superconducting supercollider are concerning and confusing to the voter, and the perceived failure of projects like the Hubble Telescope decrease confidence in our ability to achieve results even when substantial funding is provided. In spite of vehement arguments by our scientific community, congress staunchly supported the space station, and much of its funding will inevitably come from scientific budgets. More importantly, the erosion of our manufacturing and technological bases that served as a draw upon scientific ingenuity has undercut the American conviction that a strong yet chaotic science base will lead inevitably to economic prosperity. We are almost afraid of promoting scientific advance for fear that other nations will commercially exploit it! Recent congressional focus upon what we in the United States call indirect costs or overhead, and well publicized cases of questionable integrity in the practice of science, also have removed much of the shine from our armor. To top it all off, complaints by scientists about their problems in finding adequate funding and the ever more public debate on priorities and big versus little science are frequently seen as whining by a populace that has many pressing educational and economic problems.

Ocean science is somewhat protected from the worst of these problems. We are a coastal country with a maritime tradition, and the importance of the oceans to our nation is quite generally appreciated. We receive a very healthy percentage of NSF's total research support, and ocean science is a "core competency" — meaning of fundamental importance — in the Office of Naval Research, our field's second largest sponsor. Many other federal agencies — NOAA, EPA, USGS, NASA, enen the National Institute of Health — have strong oceanic interests and thus there is a diversity of sources for our scientists to tap. Our coastal states all have marine-related programs, and their representatives in Washington constitute a strong lobby for attention to the seas. While many of the states' and federal mission agencies' programs are more applied than basic, they call attention to the importance of knowledge about ocean process and characteristics.

Summing all this up, I believe that the challenge to us today is to rapidly — almost on a war footing — react to global and national needs for knowledge about the environment and man's effect on it, while evolving mechanisms for the support of our science that enhance the intellectual freedom and innovativeness of the individual researchers. We at WHOI have a good personnel and structural base from which to address this challenge, and have given considerable thought to it. I will discuss four aspects of the plans we are developing: our basic focus, our particular skills, and the roles we see for ourselves in the United States and in the international science community.

Our forte always has been the study of ocean process and characteristics based on measurements at sea. We intend to retain this focus, and increase our capabilities for both observation and analysis. We see our role as providing the basic knowledge upon which wise decisions are made and policies based, and thus work to create an intellectual environment that enhances the somewhat chaotic, but disciplined process of basic science. We recognize that education is an intrinsic part of our research, and will concentrate on these portions of the education process most closely attuned to our research focus, namely graduate and post doctoral levels.

While this may seem quite straightforward, our tight focus on the oceans and seagoing means that we will not commit our limited resources to building major computational, or space, or climatic centers of expertise, but instead will seek to expand our ability to work continuously and in all areas of the ocean. For this reason our deep submergence capabilities, for example, are very important to us, and we will move from operating separate manned and unmanned projects to developing an integrated deep ocean study system on board KNORR, which we will convert for this purpose. This plan formed the basis for our successful proposal to the Navy to operate their new research vessel AGOR-25, which should commence operations around 1997, when ATLANTIS II goes out of service.

Our commitment to improved access to the seas also has led us to highlight the Arctic and high sea states. We now have very limited abilities to work in these areas; the failure of the United States Coast Guard icebreaker Polar Sea to complete this year's Arctic expedition sharpened our concern with this deficiency, and sparked our commitment to improve. Likewise, we have argued strongly and consistently for introduction of new ship technologies like SWATH (small waterline area twin hull) that will enable us to conduct science in high lattitudes in winter, and to work through rather than run from storms. In pursuit of these capabilities, as in deep submergence, I should commend the leadership displayed by our colleagues at JAMSTEC and express WHOI's deep appreciation for their support and assistance.

Even more important than the seagoing facilities, however, is the need to improve the climate for venturesome and innovative science. Some of the joy and creativity of our work is lost when the federal funding mechanisms and initiatives place excessive burdens on the scientists — not only to propose, but to review and to manage — and when we have little or no control over the assets we operate in support of science. The only solution to such problems is independent funding, and so we — like almost all other United States educational and scientific institutions — are heavily involved in development efforts to increase our endowment. At WHOI, our objective is not only to decrease the basic burden of raising salary support on the individual investigator, but also to offer opportunities for risk that are unsupportable by our federal funding structure.

We have compared these general goals against our expertise base, and have selected a few areas where we believe special institutional commitment is needed. These efforts are intended to complement our normal process of recruitment within the departments, and indeed are oriented more to improving the environment and structure for certain critical activities than to attracting any targeted set of individuals; this remains a departmental function. The area in which we have made most progress is instrumentation. Improved access to the oceans implies a need for new measurements, and rapid technological developments combined with scientific insight have created opportunities for dramatic developments in areas as diverse as telemetry, floats, and measurement of basic physical and chemical properties. International interest in Global Ocean Observing Systems to help us understand environmental change provides additional motivation for development of a new generation of unattended instrumentation for operation throughout the world oceans. Merely one example I can highlight is the Autonomous Benthic Explorer (ABE), a joint project of WHOI and JAMSTEC. ABE will be capable of conducting precisely repeatable observational tracks at abyssal depths for periods of up to a year, allowing us for the first time to observe temporal variations in benthic activity.

To improved the environment for instrument development we introduced career ladders for engineers and information processors to parallel our traditional research track. In less than a year we have raised over $3 m from philanthropic sources to initiate innovative developments and to procure modern support hardware. We are now reviewing our structure for transitioning good new ideas into production.

A second area of historical strength we wish to enhance is air-sea interaction. Boundary layer meteorology, near-surface ocean processes and the exchange of heat, momentum and chemicals are of interest to all of our departments. Several brilliant junior scientists have recently joined us, and we are evaluating ways to capitalize on their contribution to our existing expertise. Air-sea interaction is of obvious importance to the global environment, and our efforts to build strengths in these areas are representative of our overall approach to the issues of global change, namely to concentrate on understanding the processes and characteristics that ultimately determine the climate.

The third area where we are making a significant change is marine policy. We have had a strong program in this field for nearly 20 years, but have lacked a critical mass of resident scientific staff. We have now moved the three researchers in our Marine Policy Center onto the regular scientific staff, and are commencing a search for economists and lawyers whose scholarship and interest in oceanic issues complement the research and abilities of the rest of our science community.

Finally, since I am an administrator and not a researcher, I can't resist noting that we have been making slow and careful adjustments to our management to enhance our ability to support our science, engineerig, and education. Our Associate Director of Education is now a full-time position in view of our expanded internal activities and outreach; we have expanded our development and ocmmunications staff to help us improve our funding diversity and scientific flexibility; we have added an Arctic coordinator and are more active nationally in issues involving research ships; and we have tried to reduce the administrative burden on our department chairmen. In all these efforts, a primary goal has been to keep our indirect costs low; we have one of the lowest overheads among research institutes and universities in the United States, essential to keep our total costs down since as a "soft money" institution we are so dependent upon competitive success in the federal market.

A consistent theme in United States ocean science for the last couple of years has been partnership. As our national security interests broaden to include economics, education, and environmental security, there is an increased need for institutions such as WHOI to define their relationship with the responsible government agencies. We have worked comfortably for over 40 years with NSF and ONR, and our field has benefitted from their differences in style, procedure, and interest. While we expect them to remain dominant, we must also find ways to meet the needs of other customers if they are to support our basic science.

These new demands for partnership start at a very basic and local level. While we have always tried to be good citizens, increased concern in the United States over educational quality has placed pressure on research institutions like WHOI to contribute at the K to 12 and college, as well as graduate levels. We have become involved in teaching partnerships at the village, town, county, and state levels, as well as assisting other private foundations and centers in their educational efforts. The challenge for us is to be as supportive as possible without excessively burdening the scientific staff, and in particular not to be tempted to undertake activities for which we are not qualified.

At WHOI we see similar pulls on our time and talent from the economic sector. We are a natural incubator for marine instrumentation and biotechnology products; many of our scientists have even started their own companies. In a time of local and national economic problems, we are seen as a resource. As in education, we are struggling to improve our ability to support such legitimate societal needs without impacting on our fundamental objective, to do basic research.

Of most concern to us, however, is how to interact with federal mission agencies that have increasing ocean science interests. Foremost among these is our National Oceanic and Atmospheric Administration, NOAA, which is taking a major role in the United States in addressing issues of global change. In addition to significant programs in almost all aspects of our science — TOGA, ACCP, NURP and Sea Grant are examples whose acronyms or names may be familiar to you — NOAA has two major initiatives on the drawing board which are of fundamental long term importance to our field: the Global Ocean Observing System, and a major ship modernization program. The academic community does not have a history of interaction with NOAA comparable to that with NSF and ONR; and it will be vital to us to establish one. As I mentioned earlier, similar issues pertain in our relationship with NASA, USGS, EPA, DOE and other mission agencies.

Since my colleague from Scripps will discuss the issue of new partnerships in more detail, let me merely state my primary concern. All of the United States mission agencies have sizeable in-house laboratories and research centers. In times of tight funding, there is a natural tendency to support such in-house structures first, and thus to contract less and less work to the private or commercial sector. Yet, in our field, the fundamental expertise lies in academia and institutions like WHOI. There is thus cause for concern when the sponsor simultaneously is the competitor, particularly when the national climate for science is, as I expressed earlier, undergoing change.

This challenge will be a difficult one for WHOI. Our strength lies in our independence, both from the government and from any university. Partnerships we seek must exploit this strength, not weaken it. One advantage we have is the presence in our small village of MBL, NOAA's Northeast Fisheries Division, and USGS's Atlantic Branch Headquarters. We already share many common resources such as our world class library and our computing and networking center, and hope to expand upon such facilities as well as our common research interests to establish comparative ventures.

One particularly important aspect of our independence is our ability to ignore political borders. This is important in enhancing our ability to support scientific interchange in local regional studies such as the Gulf of Maine and New York Bight, which are bordered by several states. Even on a very local level, as in

Buzzard's Bay, our independence helps us focus attention on the scientific issues irrespective of jurisdictional concerns.

More significantly, however, since our focus is on the global ocean, we value our ability to work closely on an institutional basis with our colleagues overseas. Oceanography is truly an international science, and our individual researchers historically have had very close ties with their colleagues from all nations. The multinational character of our scientific staff and students assists in those connections.

We at WHOI can, however, do more than simply encourage and passively support such natural interaction. We are able, because of our independence, to actively support both global and regional initiatives in virtually any part of the world; and because of the breadth of expertise of our scientific staff we can bring both capability and interest to bear on a very wide range of marine questions.

Two of my personal interests have been in our interactions with Japan and the nations of the CIS; and I believe our joint agreements with JAMSTEC and the PP Shirshov Institute of Oceanology have proven beneficial to both sides. My own personal interests are no more important than those of any of our staff however, and these have led to joint and regional progress with oceanographic institutions in virtually all the nations represented here, and many others.

A burgeoning interest is in the problems of semi-enclosed or mediterranean seas. While the extensive United States borders on the Atlantic and Pacific historically have focused our attention on coastal and world ocean studies, we are beginning to take a stronger interest in both common and unique issues of basins such as the Arctic, Red, Mediterranean, Caribbean, and Black Seas, and the string of seas along the Asian mainland coast. As an example of our activity in such seas, we have recently coordinated a 5-ship multinational survey of the Black Sea, followed by an international workshop held in Varna, Bulgaria, where with steering committees from each of the Black Sea riparian nations we outlined a multi-year set of national and international studies of that beleaguered basin.

This type of international activity is of particular importance to WHOI because of the opportunities it provides for access to areas, ideas, problems, people, and facilities unavailable through our normal national procedures. As good and open as our United States system is, it affords only a limited number of paths for scientific inquiry and inevitably imposes administrative burdens. Opportunities for sailing, diving, and working with our international colleagues greatly expand our options, and we plan on enhancing them.

In conclusion, we at WHOI believe our basic skills and approach — to do the best ocean science, to concentrate on making measurements at sea and under-standing them, and to support the international community — are a good match to both the scientific opportunities we envision, and the societal needs for our products: basic knowledge, and the next generation of skilled scientists. I look forward to the joys of pursuing ocean science with my colleagues from the other great institutions represented here.

Scripps Institution of Oceanography — Present and Future

Edward A. Frieman[1]

Key words. Scripps Institution of Oceanography — Oceanographic research — Global climate change — Ocean-atmosphere system — Global initiatives — World Climate Research Program — Large-scale oceanographic programs — Global ocean observing systems — International cooperation

Summary. Changing geopolitical structures, environmental concerns, economic competition in the international arena and the debate on energy and resources are all factors which affect the future of marine science. Scripps Institution of Oceanography scientists are aware of these factors and are increasingly concerned with questions of global change and the role the oceanss' play in this process. Scripps' scientists are actively involved in planning and implementing global scale ocean observation programs. Part of Scripps' future thrust will be to encourage collaboration and forge links among marine institutions of various Pacific rim countries. This interaction is considered to be vital to global scale ocean research efforts.

Introduction

Providing a snapshot in time of today's Scripps Institution of Oceanography is fairly easy. This is something that we are very familiar with and very proud of. At Scripps, the overall institutional direction is one of encouraging world class research performed to the highest standards, encouraging innovations, promoting interdisciplinary interactions, and providing access to the latest technology and/ or the environment in which to develop it. To discuss the future of the institution is quite another matter, since it requires knowing what opportunities will exist and the trends that will develop from individual scientists following their imaginations and intuitions as to the most productive line of research to pursue.

[1] Scripps Institution of Oceanography, University of California, San Diego, 9500 Gilman Drive, La Jolla, CA 92093-0210, USA

The equation governing the health of research in the United States has been good thus far, and we do not suspect precipitous change. However, changing geopolitical structures, a further awakening of environmental consciousness, economic competition in the international arena, and the emerging debate on energy and resources are also factors which affect the future of science. Given these factors and recognizing the changing nature and interests of the world as a whole, there is ample reason to believe that Scripps and other major oceanographic institutions will be increasingly recognized as major contributors to many governmental decisions. We must be cognizant of these factors, and consider the implications which they have on planning for the future of our institutions. Such changes make it increasingly important (even necessary) to work cooperatively together, and to consider the role we will collectively play in the future of the environment of planet Earth.

Key among the driving science issues for oceanographic research institutions is global change and its effect on climate and thus our environment. World leaders are taking an increased interest in the economic and social implications of global environmental changes. There is concern that changes or potential changes in our atmosphere brought about by our technological activities and resultant life styles may alter the world's environment in significant ways. This increased interest is reflected in the need by world governments for answers to basic questions related to global change processes. In my country it is also having a very immediate and direct effect on the funding in science and the scope and type of research programs being sponsored.

The Intergovernmental Panel on Climate Change (IPCC), a part of the United Nations, as well as the United States Global Change Research Program (an interagency program), have ranked the oceans and their role in global change processes, and the function of clouds and their feedback on the earth's radiation budget as the two highest priority research issues for the scientific assessment of climate change. The oceans control the timing and regional patterns of global change processes, and the clouds affect the magnitude of the change by feedback effects which determine, in part, the earth's radiation budget (how much heat is retained by the oceans and the solid earth or radiated back out to space). They are closely linked, and it is virtually impossible to study one independently of the other.

We at Scripps believe it is possible and correct to utilize basic scientific research to address major world concerns. We see no reason why research designed to elucidate basic processes cannot be applied to questions of immediate, practical concern, if the limitations of new research discoveries are made known to decision makers. Thus, a major part of Scripps' strategic plan for the future is to emphasize the dynamics of the ocean/atmosphere linkage relative to global change processes. These two high priority global change research areas are within the purview of Scripps' research, and we expect to make major progress in these areas in the future.

This paper will provide information of the development of Scripps and its present make-up, but will concentrate on discussing changes in research directions using our major thrust into ocean/atmosphere-linked processes and global change research initiatives as the prime example. Interest in global climate change and research on the various oceanic and atmospheric processes is having a positive effect on international scientific cooperation and this aspect of future research will also be discussed.

Historical Perspective — Past to Present of Scripps

The Scripps Institution began in 1903 as a biological station with private funds from local marine enthusiasts. Broadening of coverage to other disciplines started in 1908 with the employment of a staff physicist to study waves, currents, and other environmental parameters. In 1912, the institution became part of the University of California as the Scripps Institution for Biological Research. After Thomas Wayland Vaughan became director in 1924, a decision was made to broaden the fields of study to cover all marine science. The institution's name was changed in 1926 to the Scripps Institution of Oceanography (SIO) to reflect the new directions.

The transformation of the Scripps Institution into a modern, world-ranging, broadly based center of research and teaching in the earth and marine sciences really took off in 1936 when Harald Ulrik Sverdrup became director. Sverdrup was a world-class geophysicist, as well as a leading member of the school of ocean and atmospheric scientists from Norway. Shortly after he arrived, Scripps obtained its first, real seagoing vessel. After modifications to the ship, it became possible to conduct long expeditions, including two to the Gulf of California and several systematic surveys of the currents and water masses off the coasts of California, Oregon, and Baja California.

World War II inadvertently provided another boost to oceanographic research in the United States. Scripps, although small by present standards, was the largest grouping of "ocean experts" in the United States, and was called upon to provide scientific support for defense projects. Sverdrup and his student, Walter Munk, concentrated on developing a forecasting system for ocean swell and surf in shallow water, which was used to predict landing conditions on beaches planned for invasions. The temporary University of California Division of War Research, which consisted mostly of Scripps' scientists, concentrated largely on the technical problems of improving United States sonar operations. After the war, this basic research group became the Marine Physical Laboratory, SIO, which has continued working primarily on underwater acoustics and the geophysical environment affecting acoustics. The main result of the oceanographic contributions to the war effort was to help the United States government realize that it could and should support basic oceanographic research. This led to a virtual revolution in the magnitude and quality of the funding available for both Scripps and the Woods Hole Oceanographic Institution, and helped to foster the development of many other university-based oceanographic research institutions.

Without belaboring the exact evolution of Scripps since then, suffice it to say that there has been a dramatic increase in our coverage of ocean sciences that now includes strong components of earth- and ocean-related atmospheric sciences. We consider that the basic mission of Scripps Institution of Oceanography is to study, understand, and pass on to others through education knowledge of the earth's interacting systems using the oceans as the unifying theme.

Scripps is unique among oceanographic research institutions in that we have major scientific strengths in areas typically not thought to be encompassed by an oceanographic research institution. Research at Scripps involves studies of the solid earth, oceans, atmosphere, cryoshpere, and biosphere, as well as man's influence on these elements. This circumstance provides unique opportunities for multidisciplinary research as well as the basis for choosing among a number of exciting alternatives for strategic planning for the future.

Currently Scripps is conducting the most concentrated faculty recruitment campaign in our history. The academic additions to Scripps, in both faculty and research positions, represent our strong commitment to intellectual excellence, and will help set the stage for the research directions of the early years of the next century. This recruitment effort is benefitting every research division and unit at SIO. We have moved heavily into some areas of national importance, such as global change research, but we are also continuing to maintain a presence in other research areas, even though the federal support is weaker, in the firm belief that they are of true scientific interest and benefit. During the past 3 years, 20 new scientists have joined our faculty, and we are presently recruiting for some 10 additional positions. At this time, Scripps has 77 faculty members, 119 academic researchers, and 49 postdoctoral fellows.

Scripps is comprised of the SIO Graduate Department (the educational unit of Scripps), five research divisions (Climate Research Division, Geological Research Division, Marine Biology Research Division, Marine Research Division, and Physical Oceanography Research Division), two special research units (Marine Life Research Group and Center for Coastal Studies), two research laboratories (Marine Physical Laboratory and Physiological Research Laboratory), and branches of three University of California institutes (institute of Geophysics and Planetary Physics, Institute of Marine Resources, and California Space Institute). The SIO Library has been designated the Principal Collection in the Marine Sciences for the University of California system and is considered integral to the success of the research program. Scripps has a strong graduate program consisting on average of 200 students, including many from other countries.

Two new major education/research efforts are being initiated at Scripps: a Marine Biomedical Research Institute and an undergraduate teaching program in Earth Sciences. Each of these academic programs will be a positive addition to SIO and will foster interdisciplinary exchange, not only within Scripps, but with the main campus of UCSD and other parts of the marine biology community. The Marine Biomedical Research Institute will be a joint program with the

University of California, San Diego (UCSD) School of Medicine. As initially conceived, it will involve a nucleus of marine biologists, pharmacologists, microbiologists, physiologists, marine chemists and neurophysiologists from Scripps, the UCSD School of Medicine and other related scientific departments at UCSD. The fundamental goal of the program is to establish a highly interactive research and teaching environment which exploits access to marine environments for biomedical research.

Increasing public and governmental recognition of the importance of the earth sciences, coupled with the strong support by the leadership of UCSD and Scripps, have lead to plans for the creation of an undergraduate Department of Earth Sciences. Such a program will benefit Scripps by enhancing scientific interactions between SIO and UCSD and increase the overall quality of education.

Institutionally, Scripps is much more than a collection of individual research and teaching units; Scripps has a cohesiveness and functionality that enables it to perform as an organized whole in meeting its basic mission. These separate units are unified in the study of the basic processes that will provide an understanding of how the marine and earth systems function as a whole. The diversity in disciplines provides opportunities for interdisciplinary research which increasingly is a part of individual research programs at SIO. The emphasis on instruction and research and the quality of these endeavors are the primary strengths at Scripps. These strengths provide the basis for attracting the high caliber of graduate students for which we are known. The combination of the large academic and scientific support staff [492 academic personnel (includes research and teaching assistants), 500 technical support, 205 administrative support, and 66 ships' support personnel] and extensive facilities at Scripps provide an extraordinary opportunity for staff and students to have contact with and actively participate in many fields of research.

The Scripps campus itself is located on 69 ha of shorefront property in La Jolla, California. Having evolved from its original building, the George H. Scripps Memorial Laboratory (erected in 1910), it now consists of 65 buildings comprising 70,000 gross m^2 of space. In addition to housing the various departments and divisions of Scripps, they contain several specialized facilities. Among these is the Analytical Facility consisting of 19 separate laboratories, which provides access, training, and professional operation of a wide complement of state-of-the-art analytical equipment, including such equipment as transmission and scanning electron microscope, superconducting nuclear magnetic resonance spectrometer, gas chromatograph/mass spectrometer, and induction coupled plasma/mass spectrometer. Another valuable campus facility is the Hydraulics Laboratory. It is designed to provide test facilities (e.g., flow channels, wave and tidal basins, temperature and pressure calibration facilities) for simulating near shore to deep-sea conditions used in performing controlled experiments or for testing and calibrating equipment.

The development and flourishing of oceanographic research and the expansion of institutions like Scripps has been due, in large part, to the financial support given to scientific research by various individual agencies within the United

States government and also support which we receive at the state level. The majority of the State of California financial support is for the Scripps academic staff in the form of state salaried positions and to our seagoing program through providing substantial ship support funds. Scripps expenditures last year (FY91) were in excess of $72.5 million. Of these funds, 58% were from federal agencies, 24% were from the State of California, 12% from private gifts and grants, and 6% miscellaneous. The breakdown among federal agencies was 38% from the National Science Foundation, 35% from the Department of the Navy (primarily the Office of Naval Research), 6% from National Oceanographic and Atmospheric Administration, 6% from National Aeronautic and Space Administration, and 15% from other agencies.

Scripps Seagoing Operations

The Scripps Institution of Oceanography has operated both large and smaller world-ranging ocean science research ships for decades; ships have been operated by Scripps since 1908, and continuously since 1925. Scripps currently operates four major oceanographic ships and has won the competition to manage and operate the next new large research ship to be introduced into the United States Academic Fleet (AGOR 24). This ship will be 84 m and will provide a significant increase in our blue-water research capabilities. We expect to take delivery in 1994. Our current flagship, the R/V Melville is undergoing a mid-life refit, which includes new engines, thrusters, laboratories, and an additional 11 m added to her overall length to make her 85 m, as well as a new sea floor swath mapping system (the Sea Beam 2000/SIO). The R/V Thomas Washington, at 64 m, is presently very active, but is nearing the end of her useful life. The AGOR 24 will be her replacement. These ships are owned by the United States Navy. Scripps' R/V New Horizon, 52 m, is extensively used for open ocean research, but is not world ranging, and, like the R/V Robert Gordon Sproul, 38 m, is owned and operated by Scripps.

Scripps also operates two research support platforms, the Floating Instrument Platform (FLIP), at 108 m, and the Oceanographic Research Buoy (ORB), at 21 m. Both were developed under the sponsorship of the Office of Naval Research to provide extremely stable yet mobile platforms for scientific work at sea; FLIP for making accurate acoustical measurements and ORB for deploying heavy instruments (e.g., remotely operated vehicles).

Present and Future Research Directions

Recently, we have seen an increase in interagency cooperation at the federal level for funding different aspects of programs of mutual interest. This shift in emphasis has resulted in more funds being diverted to multidisciplinary, multi-institutional programs which have specific goals. The types of large research programs associated with the need to understand the possibility of global climate change are illustrative of the types of changes which are occurring in funding for the research community.

It has long been recognized that oceanic and atmospheric systems are intimately interwined. The possibility of global climate change has been the primary impetus for the mounting of major national and international research programs. These programs also provide the opportunity to acquire and/or enhance various support technologies essential for some of the new and innovative research that is developing. Below, we will discuss the involvement of Scripps' scientists in several of these major programs and give examples of some of the new research which is occurring here.

Ocean Research Programs

The sense of the scientific community and government agencies is that the time is right for dealing with the dynamical, highly coupled ocean and atmospheric systems, but that many research disciplines need to be brought together to do so. In the early 1980s, the World Climate Research Program was initiated to understand the long-term weather variability and climate change. In recognition of the central role of the ocean in these processes, the World Climate Research Program has focused on the interaction of the ocean and atmosphere. Under the umbrella of this larger program are the Tropical Ocean Global Atmosphere program (TOGA) and the World Ocean Circulation Experiment (WOCE). Many nations are involved in each of these programs, some much more so than others, but the international scope of these programs will be the major key to their success. Large programs such as these have the positive aspect of bringing together multidisciplinary (and international) groups of scientists with enough funding and time to begin to attack and get results on major questions which otherwise could not be addressed. Teams of scientists from different countries and areas of the world will be gathering complementary, needed data in all the world's oceans. The international scope of these programs also makes it easier to initiate the international cooperation necessary to obtain permission to sample in some areas. The TOGA and WOCE programs provide good examples of such programs.

TOGA

The United States Tropical Ocean Global Atmosphere program is concerned primarily with the effect of the tropical Pacific Ocean on global climate. This program has sponsored studies which will lead to prediction capabilities of the EI Niño-Southern Oscillation cycle, a dynamic climate system which affects not only the climate and ocean off western South America, but also can have dramatic effects on the weather and rainfall patterns in the United States, as well as the Indian monsoon. TOGA has established a monitoring network of drifting and moored buoys, sea level gauges, and surface ocean and meteorological measurements from volunteer observing ships in the tropical Pacific, Atlantic, and Indian Oceans.

WOCE

The World Ocean Circulation Experiment was initiated to improve understanding of ocean circulation patterns and to give the first quasi-synoptic satellite/in situ description of the global ocean. Oceanic circulation and climate are intimately entwined, and are related on a decades-to-centuries scale through the transfer of heat and momentum between the atmosphere and the ocean. The goal of WOCE is to understand the surface and subsurface circulation of the world's oceans sufficiently to model its present state, make realistic predictions of the effects various perturbations may have, and to understand the feedback mechanisms between climate and ocean circulation. Oceanic circulation and ocean dynamics are being studied using a variety of seagoing gear, but are also being estimated and studied from space with satellite altimeters.

WOCE field programs have been initiated in the Atlantic Ocean by the United States and European nations and in the Pacific Ocean by the United States and other Pacific rim countries. The main United States effort for the next few years will be to complete a comprehensive set of basin-wide observations in the Pacific and a focused set of experiments in the Atlantic. After the Pacific work is completed, at the end of 1993, the United States will turn its resources to the Indian ocean for a 1-year intensive study. The scientists present on the Scripps ship R/V Thomas Washington, from May to October 1991 for some of the initial work in the Pacific represented a truly multidisciplinary and multi-university cross section of United States ocean science. Scripps has scientists working in virtually every area of endeavor covered by WOCE. Because of the large expense of ship time, intensive sampling of hydrographic sections will be done only once, but there is a major expendible bathythermograph (XBT) program to complement and amplify the original data sets. This latter program, organized by Dean Roemmich, utilizes merchant ships and TransPacific routes as well as TOGA-funded cruises.

Sampling of specific features (e.g., the Samoan Channel) and areas of specific interest is generally being done by specific countries, reflecting that it is often more economically and/or logistically feasible for countries to work in particular geographic areas. The international framework of WOCE also makes it easier to get permission to work in specific geographical areas. For example, international cooperative agreements between France and India, and the CIS and India will make it more likely that France and the CIS will gather data from certain areas of the Indian Ocean. A cooperative program between Japan, Taiwan, and the United States will likely develop to thoroughly study the Kuroshio current. Other Japanese programs will continue their intensive, repeated study of the western Pacific.

The funding and the direction of TOGA and WOCE as well as other large national and international programs are allowing more rapid development and testing of new ideas and instrumentation than might otherwise be possible. One example of this is the drifter buoy program for measuring ocean currents and temperatures (integral to both TOGA and WOCE). It is one of the most important instrumentation concepts for physical oceanographers developed in the past 10–20 years. Traditionally, ocean currents have been measured by stationary current meters (Eulerian measurements). Lagrangian drifters, by contrast, are designed to float with the current and transmit their drift rate and sea surface temperature to satellites. Dr. Peter Niiler has significantly improved surface drifters and his instruments are being deployed primarily in the Pacific Ocean as a part of TOGA and WOCE. Over 300 drifters will be in the Pacific Ocean by 1992 and Niiler is anticipating increasing that number significantly in the near future.

Dr. Russ Davis has helped to develop a different type of Lagrangian drifter (ALACE, Autonomous LAgrangian Circulation Explorer). They can float, neutrally buoyant, at selected depths generally in the range of 600–1000 m below the surface. These buoys pop to the surface at predetermined times to transmit their data and then resubmerge. The present understanding of deep currents comes almost exclusively from density difference calculations (the accuracy of which can be questionable) rather than from direct measurements. Data from ALACE will supplement inference with direct knowledge of the currents in the interior of the ocean. Davis' work is being supported primarily by the United States WOCE program. Data from both types of drifters are vitally important for acquiring a better understanding of oceanic circulation, leading in turn, to an understanding of the linkage between global climate change and these fundamental processes. Without the funding from major programs such as TOGA and WOCE, the development and use of this innovative technology would have been significantly slowed.

JGOFS

New understandings of oceanic biology and chemistry, aided in part by new satellite techniques, have led to the development of the international Joint Global Ocean Flux Study (JGOFS). A major goal of JGOFS is to gain a better understanding of how carbon dioxide is exchanged between the atmosphere and the surface of the ocean, the amount that is incorporated by organisms and transferred to the deep sea as calcium carbonate and organic debris. Other specific objectives are: (1) to understand the global scale processes that control carbon, nitrogen, oxygen, phosphorus, and sulfur input to and removal from the ocean over time; (2) to understand how gases, salts, and water are transferred within the ocean and across the boundaries between the ocean and the atmosphere, and sea-floor sediments, and the rates of such transfers; and (3) to determine how well productivity can be measured by satellite or aircraft-borne sensors. Scripps' biological oceanographers and chemists are involved in this program in many ways.

Underwater Acoustics

Traditional physical oceanographic research as well as research related to global change phenomena is being augmented at Scripps and elsewhere by the rapid advancement in the use and applications of acoustic tomographic techniques. Sound waves in the ocean allow "visualization" of various "transparent" structures and processes within the ocean including current fields, eddies, and internal waves, as well as "solid" structures like geological features. Scripps has led the way in the development of large oceanic acoustic arrays, acoustic navigation techniques (for determining the position of oceanographic instruments), Doppler sonars for current profiling, and recently, the acoustic imaging of marine organisms. A variety of Doppler sonars developed by scientists at Scripps' Marine Physical Laboratory and used aboard our acoustic research platform FLIP have made it possible to study oceanic circulation and energetics of upper ocean physical processes in a synoptic manner not previously possible.

For many years the principal means of using sound in the ocean has been through either "active" or "passive" techniques. With an active system, an object is illuminated by a pulse of sound and its presence inferred from the echo it produces, whereas the passive approach involves simply listening for the sound that the object itself produces. A new method of employing sound in the ocean, which is neither active nor passive, relies on the natural ambient noise field for producing pictorial images of objects immersed in the ocean. This highly innovative application of water-column ocean acoustics has been designated acoustic daylight, since the ambient noise field is being used in an analogous fashion to daylight in the atmosphere for forming images. Dr. Michael Buckingham is developing an acoustic daylight, underwater imaging system, and has reached the stage where the concept has been validated through a simple experiment in the ocean off the end of Scripps' pier, which in essence provided a single pixel on a television-type screen. Methods of creating many pixels to form a complete pictorial image are currently being explored, with the ultimate aim of providing vision in the ocean simply by exploiting the natural sound illumination. Development of this capability would allow continuous sampling and integration of data on selected oceanic features.

Among the most recent imaginative uses of acoustics is the Heard Island acoustic experiment conducted by Dr. Walter Munk. The group led by Dr. Munk has plans to monitor global warming by sensing the gradual rise in the temperature of the world's oceans using acoustics. The travel time for sound in the oceans decreases with increasing temperature; thus, if the ocean is warming, the travel time for sound between fixed points would diminish as the ocean warmed. In the Heard Island experiment, large acoustic signals were sent from near this Australian possession in the southern Indian Ocean and received at various listening stations in several countries, including acoustic receivers on the United States East and West coasts, nearly 18,000 km away. The source location was unique because it provided unimpeded refracted geodesics to all five ocean basins and thus the sound signal could be heard globally. The conceptual experiment was very successful. Future experiments are now being designed which are projected to extend over at least a decade. The original experiment and future experiments are truly international, interuniversity, and interagency in scope and in cooperation.

GOOS — Global Ocean Observing System

The Second World Climate Conference held in November 1990 formally called for the formation of a Global Ocean Observing System. The conference summary statement asserted that "Climate issues reach far beyond atmospheric and oceanic sciences, affecting every aspect of life on this planet. The issues are increasingly pivotal in determining the future environmental and economic well-being. Variations of climate have profound effects on natural and managed systems, the economies of nations and the well-being of people everywhere." This Conference's further affirmation of the importance of understanding the effects of the ocean and atmosphere on the climate has clear implications on the future thrusts of oceanographic and atmospheric research.

The GOOS program itself is in the planning stages, but the basic concept behind the GOOS is that it will be a worldwide effort to combine all the various types of oceanic and atmospheric data and to both continue and initiate long-term measurement programs of physical, chemical, and biological systems in the various oceas. Long-term data sets provide the basis for recognizing cycles and, trends and discovering basic physical or biological processes and principles. There are many fundamental questions that need to be answered through a GOOS type of program. It is envisioned that such a program would have the capability of providing global synoptic pictures in near-real time which could be used to provide the initial conditions for coupled global climate models or to validate them.

Scripps is very committed to the concept of GOOS. Many of the research programs at Scripps and elsewhere, though funded through other means, would fall under the rubric of GOOS. Existing programs such as TOGA, WOCE, and JGOFS could provide much of the needed data. The drifter and XBT programs referred to earlier would almost certainly be extended as part of GOOS. Examples specific to Scripps are the CalCOFI program (California Cooperative Oceanic Fisheries Investigations) and long-term measurements of CO_2. CalCOFI is exactly the type of long-term measurement program that GOOS would encourage. CalCOFI has systematically measured the physics, chemistry, and biology of the California Current on a broad scale for over 40 years. This observational program is the largest, most intensive and extensive biogeochemical time-series ever done anywhere in the world's oceans. Another vital long-term measurement program is Dr. Charles Keeling's measurements of CO_2 started back in the late 1950s. This program is the reason why there is now an awareness of the massive anthropogenic contribution to atmospheric CO_2.

Research which Scripps' paleoclimatologists are performing would also come under the GOOS program. One of several exciting lines of research is that of Dr. Timothy Herbert. He has recently devised a study to possibly bound the range of tropical sea surface temperatures (SST) backwards in time (through at least the last ice age). High latitude temperature variability is well known, but there is still no agreement on the potential temperature range of the tropics. He is measuring lipids, which would have originated from the cell walls of cocolithophorids, in existing sediment cores from the tropical Pacific. The cell walls and the lipids (alkenones) are measurably sensitive to temperature changes of around half a degree Celsius. Such studies should yield important insights into the sensitivity of the tropical regions to major changes in earth boundary conditions over time.

A newly initiated effort which would be very important of GOOS is the Joint Institute for Marine Observations to be established at Scripps. We expect to sign the memorandum of understanding with the National Oceanographic and Atmospheric Administration very soon. We will work together to develop a joint institute focused on long-term ocean observations. The goal of the institute will be to develop a program which promotes and coordinates long-term measurement of oceanic and atmospheric parameters relevant to understanding low frequency phenomena such as global change. This institute will work on many different levels, from planning observational systems, acquiring and archiving data and disseminating it to users, to educating and training graduate students and postdoctoral fellows as well as current NOAA scientists and technicians.

Ocean/Atmosphere Research Programs

The Climate Research Division was established at Scripps 2 years ago to respond to the need for a major academic unit dedicated to the different aspects of climate research. This division is one of fastest growing at Scripps and is expected to be one of the strongest divisions in the next century. Another unit integral to ocean/atmosphere research at SIO is the California Space Institute, a multi-campus institute, headquartered at Scripps. Some of the primary research programs in the area of ocean/atmosphere research will be highlighted below. We believe these programs and their offshoots will give direction to our atmospheric research efforts well into the next century.

Scientific analyses of over 2 decades of nearly continuous satellite measurements of the earth's radiation budget data have made significant contributions to our understanding of the global mean climate and the tropical latitude heating from the sun which drives the general circulation of the oceans and other climate processes. The role of clouds is critical to these phenomena and to the earth's response to radiative heating. Certain types of clouds can increase overall heating of the planet, whereas other types produce cooling. Clouds, through their ability to retain and release heat and moisture, have an enormous impact upon oceanic and atmospheric circulations. Insight into the processes controlling cloud evolution and dissipation is necessary to produce meaningful predictions of the effects of global warming. Understanding these radiative effects, their response to climate change, and hence, their feedback effects on the oceans and the earth's climate is critical. At present, clouds and their effect on radiative heating are the largest single source of uncertainty in climate models.

Scripps is presently moving into the forefront of research on atmosphere/ocean interactions, especially the role of clouds. Dr. Richard Somerville has pioneered research in the effects of microphysical cloud processes on climate. As the greenhouse effect increases and the earth's atmosphere warms, Somerville has found that the water and ice content and particle size distribution may change, thus, changing the radiative properties of the clouds. These changes may lead to powerful feedbacks in climate through altering the ability of the clouds to reflect incoming sunlight and to contribute to the greenhouse effect by trapping heat. Dr. V. Ramanathan has found that high-altitude cirrus clouds may act like a natural thermostat for the earth, cutting off some of the sun's radiative energy to the ocean. A study conducted during the 1987 El Niño event found that more and more water vapor was pumped into the atmosphere (producing cirrus clouds) as the ocean's temperature in the equatorial Pacific neared $27°$ C. These highly reflective clouds partially shield the ocean from solar radiation, thus potentially having the effect of limiting the rise of sea surface temperature. Such research leads the way to exploring which conditions may reduce ocean heating from the buildup of atmospheric greenhouse gases. It also complements research being performed with Scripps' Climate Research Division on the causes of El Niños, the reasons for their demise, and how the transfer of heat and moisture from the upper ocean is modified and controlled by atmospheric circulation. Based in part on such research, it was recently announced that a National Science Foundation Science and Technology Center for the study of Clouds, Chemistry, and Climate will be established at Scripps under the directorship of Dr. Ramanathan.

Drs. Somerville, Ramanathan and Tim Barnett are major participants in another program devoted to study of clouds, the Atmospheric Radiation Measurement program (ARM) funded by the Department of Energy. ARM plans to attack this problem by intensively instrumenting approximately six sites in different key locations around the world. The instruments would simultaneously measure essentially all of the physical parameters associated with clouds. The ground-based measurements would be complemented by satellite remote sensing measurements and by in situ measurements taken by drone aircraft flying through and around the clouds. The drones can be thought of as low altitude geostationary satellites. This effort will continue for many years. The goal is to attempt to validate the parameterization of general circulation models (GCMs) and to increase the accuracy of predicting the effects of the buildup of CO_2 in the atmosphere. Present predictive models can differ by factors of $3-4$ because of the uncertainty regarding the effects of clouds.

The University of California's new Sequoia 2000 project is expected to provide a major boost to climate research modelling and predictive efforts by providing the next generation of needed computer hardware and software. A recent intense competition sponsored by Digital Equipment Corporation was won by the Sequoia project, a University of California, multicampus collaborative effort, and will initially provide $15 million (mostly in equipment credits) over a 3-year period. The concept of the Sequoia project is to combine computer science research with global change research in order to make major breakthroughs in computer technology. Scientists would like to be able to use large data sets located at remote sites and be able to use them collaboratively and in essence treat them as if they were small data sets. For example, scientists at Scripps might want to combine sea surface temperature data with satellite measurements, and utilize a general circulation model at the same time. The complexities of developing this capability has many ramifications, and will certainly significantly aid other efforts to model and predict the climate.

One such program which would benefit immediately is the INCOR program (the University of California's Institutional Collaborative Research Program). Scripps, together with the Los Alamos National Laboratory and the Lawrence Livermore National Laboratory, has established a joint interdisciplinary program to look at some of the coupled ocean/atmosphere processes. This program involves, among other things, a concentrated effort to couple ocean and atmosphere models in physically meaningful ways. This effort will contribute significantly to a new generation of GCMs, which study earth heating as a function of atmospheric gas concentrations. Some 35 scientists at SIO and at the laboratories have been involved. Recently, the program was expanded to include ocean and atmosphere groups at the Irvine, Los Angeles and Davis campuses of the University of California.

The oceanographic community is moving into a new era where satellite-generated data is increasingly important. For Scripps, this is especially relevant considering our strong thrust into global climate change, cloud radiative effects, and long-term ocean observations. The satellite-receiving facility at Scripps, which we have operated for the past 12 years, is being upgraded, and we will be increasing our acquisition of satellite data. Data from satellites is processed and imaged in ways that are useful to the scientific community at Scripps and elsewhere, including climatologists, oceanographers, and biologists.

The combination of satellite data with in situ oceanic measurements will provide a plethora of data that, when combined and properly integrated, will yield new insights into all areas of oceanic and atmospheric science and their interactions. For example, our researchers believe that it will be possible in the future to link measurements via underwater acoustics with satellite data in a new and dynamic simulation of the ocean and its variable states. It is likely that this balanced, two-phased attack on the problem will be the strategy of choice well into the twenty-first century.

International Cooperation and Agreements and Oceanographic Research

All of the programs and research efforts discussed above are part of Scripps' increasing involvement in ocean/atmosphere research applicable to global change questions. Without the growing concern of governments for answers to specific scientific questions of an environmental nature, we as scientists would not be able to pursue as easily such complex, multidisciplinary research. Thus, we view this emphasis as very positive. The science that is being pursued is solid, basic work that will enlarge mankind's understanding of the world around him. It is also providing an impetus to increase international scientific cooperation including having formal cooperative agreements with scientific institutions from other countries.

Scripps has active scientific interactions with scientists in many countries and has entered into cooperative agreements with other institutions. The primary purpose of such agreements is to further the development of basic research in ocean sciences through encouraging direct contact and cooperation between individual scientists, departments and institutes. Cooperative efforts between universities should be initiated by the host country and include joint research (including seagoing activities), exchange of faculty and graduate students, and exchange of information in fields of interest. As one of the major oceanographic research institutions on the Pacific Ocean, we believe that such interactions are especially important for Pacific rim countries and are exploring the possibility of entering into other cooperative agreements.

Our cooperation with Mexican universities, specifically the Instituto de Geofísica, Universidad Nacional Autonoma de México and CICESE (Centro de Investigación Científica y de Educación Superior de Ensenada, Baja California, México) provide examples of the positive effects scientific cooperation can have. Our initial interactions were started over 25 years ago and involved correcting disparities between the level of scientific programs that existed in Mexico and the United States. Mexico had not developed certain scientific curricula, programs, and technologies, but the capacity to do so was there as well as the commitment. The early exchanges involved work in geophysical geology with Fred Phleger taking the lead for Scripps, later to be joined by Russ Raitt, George Shor, John Isaacs, and Richard Schwartzlose. The level of cooperation involved Mexican students entering graduate programs in various United States universities (including Scripps), exchange of scientists to work in each other's laboratories, and field work on programs of mutual interest.

Key to international cooperation is the recognition that people work together so long as they have a common program to work on. This is not a static process and areas of interest change with time. For example, cooperative efforts with the Instituto de Geofísica started with work on basic geological formations. Presently, a state-of-the-art measurement program to determine changes in sea level, which is very valuable to global change programs, is being conducted jointly by the Instituto de Geofísica, the Center for Coastal Studies, and the Institute of Geophysics and Planetary Physics.

Early in our studies of the California Current, it became clear that the California Current system and its fisheries did not recognize international boundaries and that to properly understand the physical and biological systems and to be able to judiciously exploit the fisheries, each country needed to cooperate with the other in performing complementary studies. The United States had the California Cooperative Oceanic Fisheries Investigations (CalCOFI) program in place (run jointly by the California Department of Fish and Game, the Southwest Fisheries Center, part of the Department of Commerce, and SIO), and Mexico desired to establish a similar program. This was done in the 1970s through the Instituto National de Pesca under the

Secretaria de Pesca in Mexico. In the 1980s, MexUS Pacifico was established at the scientific working group and is the basis for present cooperative programs. The success of this program was attested to at this year's CalCOFI conference in which 15 Mexican educational institutions and laboratories participated with their United States counterparts in presenting papers and planning future work.

Exchange of scientists and students between these institutions and Scripps is ongoing, and is especially apparent at the Center for Coastal Studies and the Institute of Geophysics and Planetary Physics (IGPP). The key for this success is that the scientists meet as colleagues to collaborate on projects of mutual interest. For example a program was designed several years ago by Nan Bray and Clint Winant of the Center for Coastal Studies and José Maria Robles-Pacheco and Antoine Badan-Dangon of CICESE to work on the hydrography of the Gulf of California. This resulted in world class, pioneering work which used both United States and Mexican vessels.

The studies cited above point out several common elements in cooperative relationships with other countries. Typically there is interest in research on processes in coastal and territorial waters which will have an immediate and practical benefit to the country concerned, in programs from which there can be a high degree of information exchange, and in programs where scientists and students can receive advanced education and training in techniques not available in their country. A project recently started at Scripps, the Little Bight Study sponsored by the Minerals Management Service, is expected to generate just this sort of exchange. This project will be studying circulation patterns and shelf processes in the southern California bight and is likely to be amplified in scope in the near future. It will provide a model for studying physical processes in near shore waters and formulating a synoptic view of the interactions of these processes. The practical benefits of such a study are to better understand diffusion and dispersion in the near shore marine environment (e.g., pollution of near shore waters, coastal erosion, predicting effects of offshore mining and drilling operations).

Cooperative, international interactions are especially important for performing research associated with major scientific issues like global climate change. Whether the data being collected are from coastal or near coastal areas, the high seas or the atmosphere, only through the free exchange of ideas and the gathering of information can we as scientists truly begin to tackle such questions.

Russia

Scientific Research at the P.P. Shirshov Institute of Oceanology

Vyacheslav S. Yastrebov[1]

Key words. Oceanology — Marine biology — Ocean geology — Ocean physics — Black Sea — Ecology — Research vessels — Ocean turbulence — Ocean devices — Scientific fleet

The P.P. Shirshov Institute of Oceanology

The P.P. Shirshov Institute of Oceanology is a scientific institution for multi-disciplinary study of the ocean. Since the institute was established, there have been many examples when a multidisciplinary approach has permitted us to solve large fundamental problems, the case of the phosphorites being typical.

The staff at the institute at present number 2,200 people, including 656 scientific personnel, 606 engineers, and 550 staff members working as the r/v crews. A total of 1,100 members of staff work in moscow. The research personnel comprises 2 full members of the USSR Academy of Sciences, 7 corresponding members of the Academy of Sciences, 5 full members of the Russian Academy of Natural Sciences, and 4 corresponding members of this Academy.

The institute possesses eight ocean-going research vessels. There are several subsidary branches at the institute:

The south branch in Gelendjik with 450 staff members
The Atlantic branch in Kaliningrad with 600 staff members
The Leningrad branch with 50 staff members
The Murmansk branch with 10 staff members

The main base of the institute's fleet is in the port of Kaliningrad. Researchers at the institute are using the following technology and devices:

self-contained bottom stations for various purposes
towed instruments and vehicles

[1] P.P. Shirshov Institute of Oceanology, Russian Academy of Sciences, 23 Krasikova, II7259 Moscow, Russia

manned underwater vehicles, including deep-ocean submersibles
shipboard diver's hyperbaric units
standard oceanographic instrumentation and equipment

The subject matter of research at the institute was formed on the basis of the following principles. Priority was given to:

Multidisciplinary research
Research programs that are capable of supplying drastically new data on the ocean
Research within the framework of large international programs
Research having ecological implications

Marine Biology

The central focus in the field of marine biology is on the problems of human impact on ocean ecology. This includes study of the reaction of the ocean to the existing human-induced stress, regulation of the ocean to the existing human-induced stress, and regular monitoring of those ocean areas that manifest symptoms of a dangerous situation in terms of ocean ecology. The outflow of pollutants with rivers, along-shore currents, the ecology of the shelf and central ocean areas, the role of plankton in the cleaning of the ocean from the surface, the burying of substances in sediments, the reprocessing and transformation of matter in the near-bottom area — all these major problems are being explored.

Ocean Geology

In the field of ocean geology, the leading methodological principle in research since the late 1960s has been the lithospheric plate tectonics theory. This is a productive concept that has permitted scientists to focus their main attention on the most active zones of mid-ocean ridges. There are first of all areas of active hydrothermal manifestations, and here fundamental and applied problems come together. The fundamental aspects are concerned with fluxes of substance and energy from the interior of the earth, and evalution of their Role is of principal importance. The applied aspects are related with mineral formation in the vicinity of hydrotherms.

The hydrothermal vents in the ocean are an example of a necessary multi-disciplinary approach to a solution of a problem. The hydrotherms, as a unique object of research, are of interest to geologists, biologists, physicists, and chemists. A fantastic living world occurs here as well as the formation of valuable ores and physical interactions between the outflowing hot waters and the main mass of the ocean.

Ocean Physics

In the field of ocean physics the major interest is centered on ocean in-homogeneities of different scales: a global system of currents and mesoscale variations such as fronts, rings and water lenses. Among other phenomena are micro-inhomogeneities and turbulence. In various ocean areas, supprisingly complicated structures have been found to occur. Most interesting are turbulence fields in seamount areas, such as around Ampere seamount in the Atlantic Ocean.

One of the most interwined multidisciplinary problems is the problem of the near-bottom layer. Areas of interest include the following aspects: the isolation phenomenon, near-bottom storms, local vents of endogenous hot waters, and complex biochemical processes at and on the bottom.

Black Sea

A great deal of research at the institute is related to the fate of the Black Sea. Decreases in fish populations and huge amounts of pollutants brought by rivers have been typical features of late. The vacated ecological niches left by the fish have been filled by jellyfish. However in 1988 a new organism, ctenophore, was introduced into the Black Sea, and to a large degree this has superceded the jellyfish. Another important question is hydrogen sulfide saturating the Black Sea waters, including both deep-sea and nearsurface hydrogen sulfide.

The hydrology of the Black Sea to a large extent determines the processes of biological alterations. At present the principal task is to secure the regular monitoring of the Black Sea, to study the content of the runoff of the main rivers, and propagation of pollutants from their source. Similar tasks are also planned for the Baltic Sea.

Special Lecture

Earth's Environment and Ocean Research

PROF. DR. JIRO KONDO[1]

The Origin of the Ocean

The earth in the solar system is presumed to have been created by the collision of small planets or asteroids which measured about 10 km in diameter. Containing volatile matter such as water, carbon, and so on, these gases were suddenly released at the time of the collision and formed a primeval atmosphere to envelope the outer crust of the planet during its growth. The asteroid or small planet contained metallic iron and when the metal reacted with steam, the iron oxidized and hydrogen remained. This hydrogen and carbon reacted to create methane. During the initial stage when the primeval earth started to grow, the earth's crust is assumed to have been enveloped in an atmosphere of hydrogen, methane, steam, and so on. However, as the primeval earth grew and the gravitational force increased, the amount of energy released by collisions on the earth's surface augmented and the absorption the primeval atmosphere of their heat radiation caused the temperature of the earth's crust to rise. As a result, the metallic iron and steam reaction no longer progressed. The overall volume of steam increased and the volume of hydrogen decreased. This led to carbon becoming stabilized as carbon monoxide. Therefore, the primeval atmosphere over the high temperatures of the earth's crust which approximated the melting points of rocks, chiefly consisted of steam and carbon monoxide. The amount of steam in the atmosphere, as the earth's crust was covered with magma, stopped fluctuating in volume. The total volume approximated the volume of the oceans today. Despite the fact that 80% of the atmosphere was steam, the environs of the earth's crust were very dry due to the high temperature. The very dry convective layers went to heights of 300–350 m above the earth's crust. Higher than that, the atmosphere cooled and became moist. The moist convective layers were about 50 km thick and generated clouds from which rain fell. However, this rain did not reach levels below 300 km from the earth's crust.

---

[1] President, Science Council of Japan, Professor Emeritus, The University of Tokyo

These dry convective layers could exist in a stable state only when the earth's crust retained a high temperature. The primeval earth grew, its radius approximating the earth's radius of today, and the volume of falling asteroids decreased. The energy released on the earth's crust reduced and then as the temperature of the earth's crust decreased, it became stabilized. In other words, the dry convective layers disappeared and moist convective layers covered the earth's crust. This meant that rain reached the earth's crust. As a result, steam in the atmosphere became the ocean which how envelops the earth's crust.

The atmosphere of carbon dioxide was transformed to the present nitrogenous atmosphere when continents were formed on the earth's crust and rocks, formed by carbon dioxide which had dissolved into the oceans, were added. It can be said that the environment of the earth determined that the earth would be a planet which contained continents and oceans. Of course continents are a geographical form only existing on the earth.

Ocean Water and Ocean Currents

Ocean water comprises 98.3% of the total volume of water near the earth's crust.

The ocean in comparison to the atmosphere has 100 times the heat capacity and 270 times the mass with a total presence of carbon dioxide 52 times that of the atmosphere. Consequently the role that the oceans play on the environment is very important.

The density of salt in ocean water is on average 3.5%, with these salts being separated to ions. The ocean contains just about every known element but minute elements such as nutritive salts fluctuate considerably depending upon the place, depth, and season, directly affecting the existence of marine life.

Oxygen and carbon dioxide as dilutants in ocean water are indispensable for the respiration of living creatures and photosynthesis. The surface temperature of ocean water rises according to the proximity to the equator and decreases with proximity to the poles. Near the equator it can exceed 28°C. The higher the temperature, the lower the density of ocean water, and the differences in specific gravity with ocean water of high density give rise to the formation of ocean currents. Currents are also born from the wind and from the influence of the Coriolis effect originating from the earth rotating on its own axis. The flow of these ocean currents can now be mathematically calculated by means of fundamental equations of fluid dynamics.

Convection currents in the Pacific and Atlantic Oceans flow clockwise in the northern hemisphere and counter clockwise in the southern hemisphere. In the western portion of the oceans there are narrower and stronger currents than in the eastern portions that are known as the west coast strong currents. The Japan Current is a representative example of this type of current.

Looking at the strata of water in the oceans off the shores of Greenland in the northern Atlantic for instance, the deeper strata are caused by the low temperature high salt content water remaining after the formation of ice, sinking low into the ocean depths. This deep strata water then moves south in the Atlantic and after circling the Antarctic continent reaches the offshore area of New Zealand to change course northward and enter the northern Pacific. During this time, pressed forward from behind by further currents, this water gradually rises and then mixes with the surface ocean water of the northern Pacific and finally appears on the surface. However, this circulation takes 1,000–2,000 years, and the scale of circulation of ocean currents is in units of 1,000 years.

The dissolved oxygen in such deep strata water, while taking 1,000 years to move, gradually decreases as it is consumed in the decomposition of organic matter which has fallen from above. Therefore the older the water is, the less the amount of dissolved oxygen. At depths of around 400 m in the Pacific Ocean off the shores of Peru and California where the deep strata water finally rises, even today there is hardly and dissolved oxygen left. In ocean water where there is no dissolved oxygen, fish cannot survive. Of the chemical nutrients which are required by marine life, the most usual insufficiency is in nutritive salts. The major items are phosphorous, nitrogen, silicon, and so on, but nearly all marine life requires phosphorous and nitrogen, whereas diatomaceous matter suffices for silicon. Phosphorous and nitrogen increase with human activity but silicon is naturally depleted. Recently human activity, instead of increasing diatomaceous matter which represents the natural state of ocean water, has increased the likes of Flagellatae and Dinoflagellatae. The toxic red tides caused by *Chattonella* and *Goniolacs* which make scallops toxic are good examples.

Marine Development and Marine Pollution

In Japan, due to surging population and growing wastes resulting from industrialization, water pollution of the rivers and oceans has gradually been highlighted as a problem. The most disastrous example was the cases of Minamata disease caused by the Ariake Sea being polluted with inorganic mercury from plant effluents. With regards such pollution many factors need careful consideration, including the poor quality of water, the worsening quality of water in rivers, lakes-, and swamps, and the increasing scarcity of fish. In extreme cases, offensive smells from polluted water bothers nearby citizens, and in this way, people begin to pay attention to the destination of polluted water.

In this regard, oceanic development deserves special attention from the aspects of resources, energy, foods, space, and waste disposal. Figure 1 shows 24 development-related items set forth in the 1980 January report submitted by the Oceanic Development Council. At the center are the major fields of use of the ocean, while the black domain illustrated in the outermost rim shows the risk of environmental pollution that such development projects can cause.

Marine Resources

Marine resources can be grouped into two areas, seabed and seawater resources. Manganese, gravels, petroleum gas, and coal can be taken from seabed too.

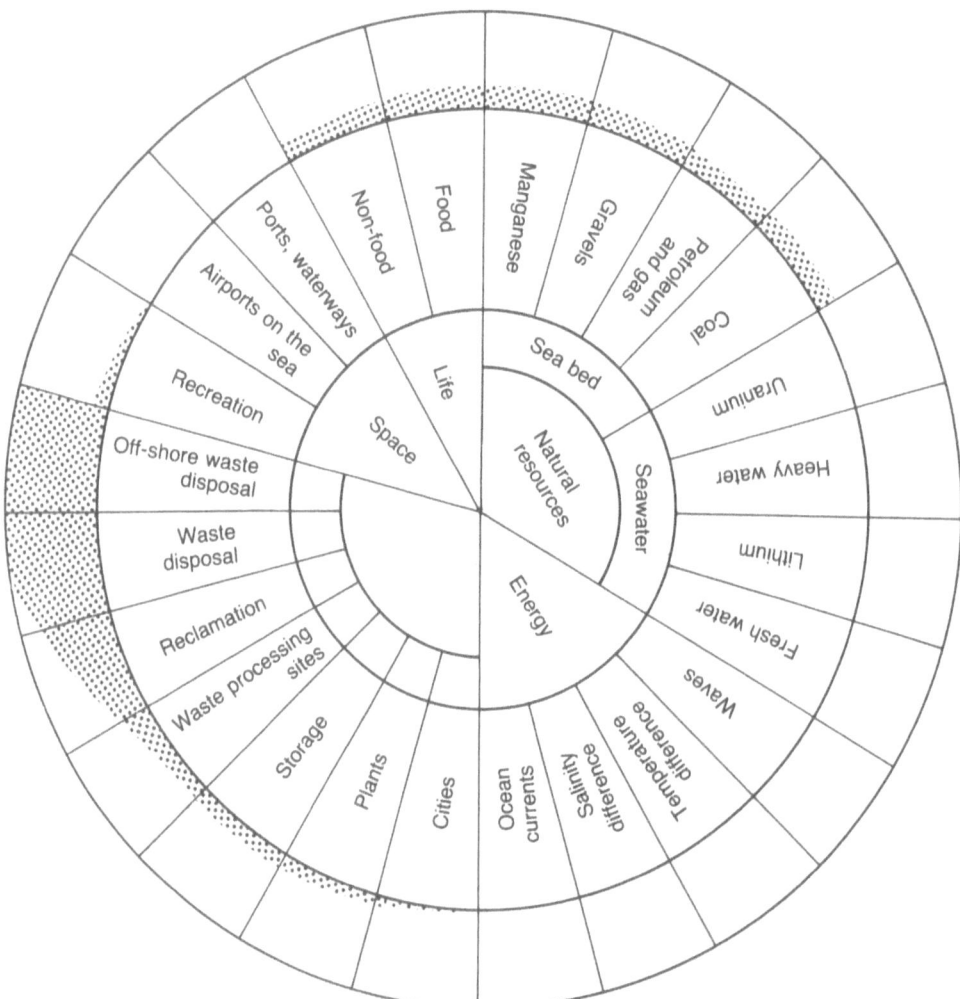

Fig. 1. Ocean utilization prospects. The *black areas* in the outermost rim show environmental pollution risks. (Redrawn from The Oceanic Development Council Report, January 1980)

Recently, the removal of gravels from shallow waters has caused drastic changes in marine ecological systems. This kind of act, if under taken without careful consideration, could pollute the ocean, which is why a thin black portion appears on the outermost rim of Fig. 1.

Resources such as uranium, heavy water, and lithium for nuclear power generation and fusion, are available in seawater. Also , desalination of seawater provides fresh water for use as domestic and industrial water. By the year 2000, five times the amount of fresh water used now will be needed, although this amount of demand is least likely to cause ocean pollution.

Oceanic Energy as New Energy Sources

To overcome an expected shortage of fossil fuels ahead, efforts must be made to use waves, thermal differential, concentration differential, currents, and so on as alternative energy forms in place of fossil fuels. Still remaining at the study stage, these renewable energy sources are not promising for immediate use, but they are the least likely to cause pollution.

Use of Ocean Space

This field involves methods to use the ocean based on the concept that the ocean provides additional space extended from land. Land reclamation is in practice in two ways, along the coast and in offshore areas shallow enough to permit such projects. For example, coal/petroleum plants can be constructed offshore or undersea, and oil storage tanks and waste disposal sites can be installed onshore or offshore. Apart from land reclamation, ocean dumping of wastes is also in practice. A plan to store spent nuclear fuels in an island in the South Pacific has been banned due to opposition by local peoples. Because these uses of the ocean often result in pollution, a thick black portion appears on the outermost rim of Fig. 1. In addition, using the oceans would double the entire land area available on the earth. Traffic space to connect lands, ports, and sea routes are being developed, and plans are under consideration to establish additional ocean transport and traffic systems, install such offshore facilities as the Kansai International Airport, and develop marine resorts for sea bathing, yachting, and so on. By 2000, marine transport is likely to almost be quintuple present levels, which could destroy ecological systems by bringing new varieties of creatures in. A good example is the coral reefs in Okinawa where corals have been seriously damaged by *Ancanthaster*.

Foods and Non-Food Products

The final section listed in Fig. 1 is the use of oceanic life as direct foods and/or non-food products. Here, the non-food products are feeds and the like, which do not provide human foods but indirectly contribute to food production. Given a food chain, one practically needs plants ten times heavier in weight than farm products per unit calorie, if protein and fats required in the diet must be obtained from land animal meat. While the Japanese have long depended on fish and shellfish as their principal protein source, positive food development efforts will be needed from now on through such systems as marine pastures. In the meantime, food resources such as krills must concurrently be developed. Also, even if not designed for human foods, harvesting feedstocks of feed among oceanic life is a very promising area. However, excess production of life in the ocean, particularly in closed waters like inland seas, can impair the balance of ecological systems, actually causing the red tide phenomena. The nitrogen and phosphorus inflows from land causes such eutrophication. While excess stockbreeding and grazing on land is alleged to change soil and cause desertification, it appears nothing but natural if the same can develop undersea. In this regard, shown in the left half of Fig. 1 are the developments that can have particular effects on the pollution of oceanic environment. Even if they would not directly pollute the ocean, these developments could destroy the ecological system and cause irreversible results, so their implementation should be paired with maximum possible caution.

Environmental Changes Caused by Development

Being a natural system, the ocean has self-cleaning functions. The total amount of water contained in the sea is so huge that some pollutants can easily be diluted with just a negligible effect, if any. However, ideal dilution requires agitation until the pollutants become uniformly distributed. Ocean currents, winds and waves perform this function to some extent, but it is now becoming obvious that they are not so powerful as to be capable of producing sufficient diffusing effects. It also can clearly be noted that pollutants discharged from ships or oil spills from tankers run aground precipitate and/or are carried to the seabed, without spreading very far. In such events, oceanic pollution can have grave effects that influence, via evaporation, not merely the ocean but also the land, where soil deterioration can be caused. Once destroyed, oceanic ecological systems cannot easily be restored. Land life was born undersea, and it would naturally be hard for today's land creatures to remain alive once the mother sea was dead. Therefore, whenever oceanic development projects are planned, it is essential to assess whether restoration to the original state is always possible. In other words, while confirming reversibility, special attention must be paid so that development should never result in an irreversible state.

To thie end, it is necessary to forecast the environmental changes caused by development, and assess such changes. For the present, three things can be cited. First, oceanic ecosystems must be studied, and secondly, seawater must be monitored. Third, hazardous wastes generated on land must be disposed of on land by establishing a closed system so that their ocean dumping can be prevented.

Oceanic Ecosystem Studies

The need for oceanic ecosystem studies concerus both the open ocean and the coast alike. Given that greater human influences are obviously produced on the coast, observation and study efforts should first be concentrated on the coast and adjoining seas. It is harder to study undersea plants and animals than their land counterparts. Given that few comprehensive study results and data appear available on plankton, fish, and shellfish in individual waters, such study efforts have to be strengthened. Also, by preparing a microcosm or a vessel to observe marine ecology, scientific studies must be made to learn any recovery potentials inherent

Fig. 2. Japan's US-1, a STOL flying boat that is capable of landing and/ or taking off under conditions such as 3 m-high waves and 25 m/s wind velocities. (From Kondo J "Why do planes fly?" Kodansha Bluebacks)

in ecological systems. There are research institutes and coastal laboratories dedicated to oceanic studies, including the Oceanographic Research Institute of the University of Tokyo and some operated by government agencies, but their number is still limited. From the standpoint of fisheries, it is pleasing to see oceanic studies in progress, but, from the standpoint of environmental science, ecological studies remain unsatisfactory. Because elementary ecological observation does not necessarily require sophisticated equipment, it is first hoped that an increasing number of researchers across the country will start studying oceanic ecosystems.

Monitoring

It has been reported that DDT was detected in whale meat from the Antarctic Ocean, so monitoring must be made by paying special attention to how the quality of seawater can change in the long run. Observation of this can also be made from space via satellites like LANDSAT and NOAA, but it appears crucial to take seawater samples directly from the seas and analyze them. Sampling methods are largely dependent on ships without alternative means, meaning that special ships must be built and fitted with equipment such as seawater sampling machines and chemical analyzers.

Japan's US-1 is a STOL flying boat equipped with a blower in the boundary layer of its main wing (Fig. 2). Also given a specially designed wave cutter, it is capable of landing on and/or taking off from seas under poor conditions such as 3 m-high waves and 25 m/s wind velocity. A new sampling method is being considered based on a flying boat of this class. The boat could be used in regular sampling from fixed points on the sea, with the samples promptly taken back to allow immediate chemical analysis. As is well known, the ocean spreads as if unlimited and, therefore, how to monitor the quality of seawater poses an important subject which requires continuous study efforts in the future.

While concentrations of pollutants in the environment are extremely low in a normal state, they can sometimes be found enriched in fish and shellfish due to the food chain, among others. Accordingly, an effective method of monitoring

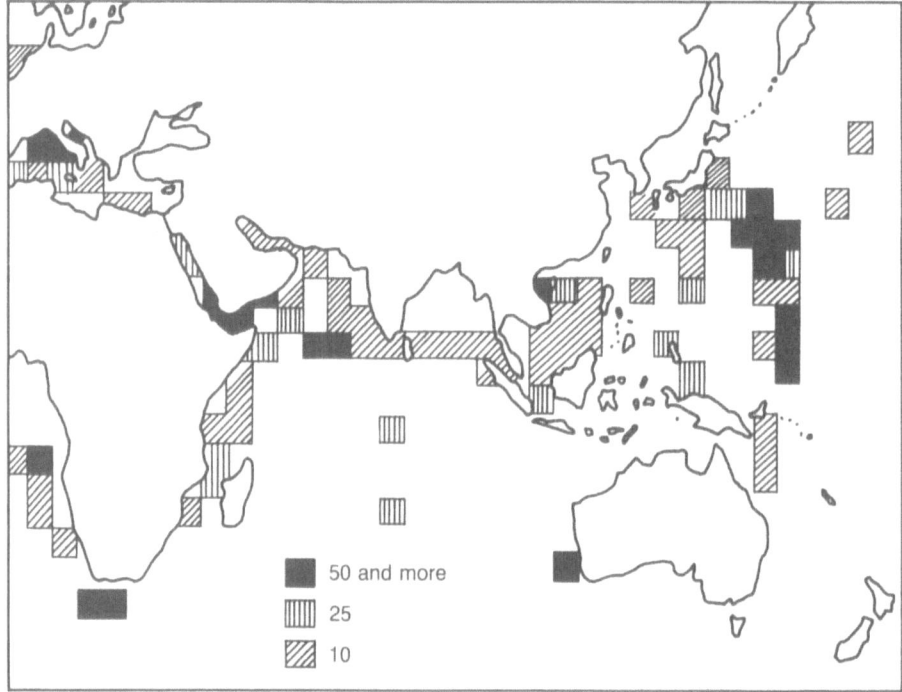

Fig. 3. An example of ocean pollution: Frequency of oil slicks found in the Eastern Hemisphere sea areas. Sea areas in or around the Japanese waters were among the worst polluted in terms of oil slicks (spills from tankers, etc.) found. *Black areas*, over 50%; *striped areas*, over 25%; *hatched areas*, over 10% (From Intergovernmental Oceanographic Commision (IOC), UNESCO)

oceanic pollution is to sample certain shellfish, such as musses, and to chemically analyze their meat and internal organs to measure intaken pollutants, particularly heavy metals and soluble substances. Environmental monitoring is routine, painstaking work, but, stepped-up efforts should be made right now, because it will be too late to initiate such efforts when pollution is obvious to everybody's eyes.

Needless to say, monitoring efforts should be made on a global basis under international cooperation. Given that many countries in the world after all are interlinked through the seas, they should initiate a system of multilateral cooperation in their efforts to prevent oceanic pollution. Just calling for environmental preservation will always end in vain so long as some imprudent transport ships continue to discharge waste oil or used washing water into the ocean. In short, it is a problem of environmental ethics, which requires peoples all over the world to behave themselves, watch each other, and commit to satisfactory monitoring. If someone sneaks an illegal ocean dumping, as done on land, just monitoring is not so helpful either. That is why an international convention on oceanic oil pollution must be agreed on at the earliest possible opportunity. Illustrated in Fig. 3 is the present situation of oceanic pollution.

Disposal of Trazardous Wastes

Despite fears that the ocean can be polluted, to keep mankind on land and leave the ocean untapped is an idea which can never be agreed on. Rather, in the years ahead, increasingly positive efforts must be made to develop the ocean to put it to effective use for the benefit of mankind. Such efforts, however, must rest on a system engineering concept or the like, where land and the seas are united as one system.

Also, needless to say, it is desirable to dispose of municipal and industrial wastes on land by establishing a closed system, which naturally involves huge capital outlays. Even if huge, the cost should be justified as rational spending to preserve environment of the only earth. On the other hand, instead of using the ocean as a waste storage yard, investing in marine environmental preservation as part of a single global system deserves serious examination. Then, what is needed is basic knowledge of oceanic environmental science. At least, prior to starting any oceanic development projects, it is necessary to identify causes of pollution and assess likely effects in the future.

Petroleum Contamination

Today the whole world turns to petroleum as a source of energy. Since areas producing petroleum in the world are confined to a limited number, petroleum is transported around the world by ocean-going tanker vessels. However if the tankers run aground and the vessels' hulls are damaged, large amounts of petroleum or derivative products spill out into the ocean. The specific gravity of petroleum is lighter than ocean water so that it spreads over the surface and moves with the action of the winds, waves, currents, and so on, becoming dispersed. During that time, changes take place through processes such as evaporation, emulsification, dispersion, dilution, precipitation, photochemical oxidation, and bacteriological decomposition.

On the 16th March 1978 the Amano Cadiz, a supertanker operating under the Liberian flag that was fully loaded with petroleum from Saudi Arabia and on its way to the UK, ran aground approximately 1 km off the shore of Brittany and 246,000 kl of petroleum cargo spilled into the ocean. The shores of this peninsula for about 200 km were contaminated. It took 2 months to recover the petroleum floating in the ocean and 6 months to complete the cleaning and elimination of oil contamination washed ashore. In 1989 the Exxon Valdez ran aground off Alaska and 41,000 kl of petroleum contaminated the shores of the surrounding bays, for a distance of about 1,750 km. In March 1991 due to the Gulf War, the oil wells of Kuwait were destroyed and petroleum leaking out of loading facilities contaminated the waters of the Persian Gulf.

Contamination of the oceans by crude oil differs a great deal according to the type of oil involved. Here I would like to review the kinds of environmental damage caused by the discharge of petroleum.

Evaporation

The volatile elements in crude oil are lost through evaporation from the surface of the ocean. Consequently the toxicity and volume of the oil is reduced accordingly. In the case of crude oil, 30%–50% evaporates into the atmosphere due to its volatility. This occurs within 1–2 days. With less volatile heavy oil and naphtha fuel oil, only about 10% will evaporate. However the lighter derivative products such as gasoline and kerosene evaporate completely.

Emulsification

Oil is churned in contaminated currents by wind and waves and mixed with the ocean water. It eventually forms an emulsion consisting of drops formed from blobs of high viscosity oil containing 60%–80% water. These make recovery from the shores and ocean surface and subsequent disposal exceedingly difficult. Most of the crude oil usually emulsifies within 2–3 weeks to become blobs of oil containing 70% water.

Dispersion

Some types of oil, when subjected to wind and waves, break up into small droplets as they spread. This is known as dispersion. When oil is dispersed in this form, its toxicity towards marine life is reduced and therefore this is beneficial in pressuring the marine environment. With specially prepared chemicals, this dispersion process can be accelerated. These small droplets can be further broken up by employing bacteria for biodegradation of the oil to make it harmless.

Solution and Sedimentation

A majority of hydrocarbons do not dissolve in water but some elements in crude oil, particularly light, low ignition point fluorohydrocarbons, quickly dissolve in ocean water. However this category are more liable to evaporate rather than dissolve. Of the floating oil droplets, those with a specific gravity heavier than water sink to the ocean floor and settle as sediment, but in actual fact, there are very few cases in which oil is transformed into material of such heavy specific gravity.

Photochemical Oxidation

When oil is bathed in sunshine on the ocean surface, oxidation commences as a result of photochemical oxidation. As a result, a portion of the elements decompose in the ocean. As the water soluble portion increases, the oil is effectively dispersed through dissolving and dilution. The speed at which photochemical oxidation of the oil progresses is greatest when the oil is no the surface of the ocean or has been washed upon the shore and entirely exposed to the atmosphere.

Biodegradation by Bacteria

Decomposition by bacteria is the most important process in considering the ultimate disposition of oil in a marine environment. The speed of decomposition differs according to the marine environment but it is said that bacteria contained in $1 l^3$ of ocean water can decompose 30–350 grams of oil in 1 year.

Of the aforementioned processes, the most devastating to the ecosystem is said to be the emulsified oil that is highly viscose. When this is washed up on the seashore, it becomes fatal if it coats shellfish and is highly damaging if it contacts birds' feathers. Up to the present time, oil slicks have been observed at all of the places marked on Fig. 3 and every one of them is located in a recognized oil tanker shiplane. In the case of the Exxon Valdez tanker in 1989, a settlement was attempted by the payment of US $1.1 billion in damages but the United States Federal Government refused. Again the oil contamination of the Persian Gulf by Iraq is being internationally criticized as an act of environmental terrorism.

The Valdez Principles refer to the Exxon Valdez oil tanker which contaminated over 2,000 km of shoreline and killed over 450,000 ocean birds. The United States environmental protection group CERES proposed ten principles that corporate industrial organizations should govern their actions by and gave them this name, the Valdez Principles.

Red Tide/Blue Tide

In the ocean, particularly closed off bodies of water such as bays and ports or inland seas and lake swamps, the generation of bacteria and laver in response to the inflow from the land of nitrogen, phosphorous, and so on is called eutrophication. In the case of lake swamps this process is called AOKO or fresh water red tide and in the case of ocean water, red tide, blue tide, and so on.

In the case of AOKO, anil laver sives the water surface a greenish hue. When the laver is decomposed by microorganisms, a detestable odor develops and oxygen is consumed so that there is a lack of oxygen for fish that are being raised on fish farms. In the red tides, plankton in the ocean water increases rapidly dyeing the water red because of the volume of plankton and laver. In this case also, fish are critically affected. The generation of red tides in the Inland Sea are well known for the critical damage they have caused on mackerel or yellow tail being cultivated on fish farms. The same phenomenon has also been confirmed in closed bodies of water in many other parts of the world. The blue tides are caused by advanced organic contamination and eutrophication at the bottom of closed bodies of water creating a shortage of oxygen which results in oxygen-short clumps of growth rising from the bottom to the surface and siving the surface a bluish hue. Like the red tides, this causes massive quantities of fish to die of lack of oxygen.

Sea Lions Killed in the Baltic

From around 1988, 50,000–100,000 sea lions along the shores of Sweden and Denmark began being affected by the contamination of the ocean. Autopsies of the animals revealed that the lungs, eyes, stomachs, and intestines were inflamed. By February 1989, 17,000 of them had died mainly due to a contagious virus. Fortunately no further deaths have been reported but the 15 rivers that flow from the European continent into the North Sea are said to be the carriers of a large volume of toxic substances, including heavy metals which are the contaminants responsible for the deaths of the animals. The animals were weakened by these contaminants which indirectly led to their ultimate deaths.

In any case the large number of deaths of these large mammals each weighing over a ton made the northern European peoples very much aware of the importance of avoiding or guarding against ocean contamination. Amongst the porpoises of the Pacific Ocean and the North Sea, the presence of PCB has been confirmed and the density has reached several tens of parts per million. This has been concentrated several million times within the internal organs of the porpoises. PCB began to be used in 1929 and in 1965 was found to be also present in the bodies of whales.

DDT and BHC are often used as a pesticide against the mosquito conveying tropical malaria. These substances are carcinogenic and the manufacture and use of these substances has been prohibited, but in the developing countries they are still being manufactured. Exceedingly dangerous materials such as dioxins have also been discovered. However the presence of heavy metals, DDT, BHC, PCB, and so on have been clearly confirmed as a result of analyses of over 150 different substances. It is estimated that excrement of over 150 million Europeans is carried from the continent by the rivers into the North Sea. This is the equivalent of 100 million tons in a single year.

Encroaching Upon the Oceans and the Atmosphere

Contaminants dissolving into the oceans again evaporate from the oceans to reenter the atmosphere. As the oceans and atmosphere are so large, it is impossible to grasp in a numerically definite volume sense their mutual relationship. Furthermore it is assumed that there are differences arising according to the changes that are possible in the ocean surfaces and currents. However to solve this problem scientifically is very important for determining policies to be adopted regarding problems related to global warming.

In constructing the molecular structure of the carbon atom, the circulation of carbon does not always come out evenly. Burning of fossil fuels by conversion of carbon indicates that in a year 20 billion tons of CO_2 are released into the atmosphere, of which the residual amount remaining in the atmosphere is less than almost half of that amount. Today when destruction of the forests is proceeding, no other absorbent like the ocean is conceivable. It is not yet clear how CO_2 is absorbed into the ocean. Vegetative plankton and algae in the ocean, like land vegetation, fix CO_2 through photosynthesis. By cultivation fish and shellfish which feed on this, and CO_2 is changed to lime and $CaCO_3$. If this is again dissolved in the ocean, nutritious salts such as nitrogen and phosphorous together with minerals like iron and so on dissolve into the ocean as available nutritious salts.

If the $CaCO_3$ of fish could be considered the intermediary, ocean water could once again absorb large quantities of CO_2. This is the research project of Kikuo Kato of the Institute of Aquascience of Nagoya University.

Panel Discussions

Research in Deep Seas

Panelists: CRAIG E. DORMAN, LUCIEN LAUBIER,
VYACHESLAV S. YASTREBOV, KAZUO KOBAYASHI, ISAO KARUBE,
HIROYUKI NAKATO
Session Chairman: NORIYUKI NASU
Rapporteur: HIROSHI HOTTA

KAORU MAMIYA: We would like to start the last session for today, the panel discussion session on the topic of "Research in Deep Seas."

I would like to introduce Professor Noriyuki Nasu, who is professor of the University of Air, and professor emeritus of the University of Tokyo, Japan. He served as the director-general of the Ocean Research Institute of the University of Tokyo and thus contributed to the advancement of oceanography. He is also the chairman of the Council for Ocean Development, an advisory Committee to the Prime Minister, and a Japan representative of UNESCO's Intergovernmental Oceanographic Commission (IOC). Before we start the panel discussion on deep-sea surveys and research, I would like to introduce the six panelists.

The first is Dr. Craig Dorman. Dr. Dorman is the director of Woods Hole Oceanographic Institution, and his specialty is physical oceanography. Next is Dr. Lucien Laubier. He is director of IFREMER's International Pelations and Cooperation, and his specialty is marine biology. Next to him is Dr. Vyacheslav S. Yastrebov. He is the director of the Shirshov Institute of Oceanology in Russia. Oceanography is his specialty.

Professor Kazuo Kobayashi, is professor at the Ocean Research Institute of the University of Tokyo. Geophysics, especially that of the deep sea bottom is his specialty. Next to him is Professor Isao Karube. Bioelectronics is his specialty, and he is professor at the Finontier Science and Technology Research Center of the University of Tokyo. Finally there is Mr. Hiroyuki Nakato, executive director of JAMSTEC, and naval architecture and instrumentation are his specialties.

This concludes my very brief introduction of the panelists. Another important person I have to introduce to you is Dr. Hiroshi Hotta, director of the Deep-sea Research Department of JAMSTEC, who will serve as the rapporteur of this session.

As a background, I would like to say the following: Global-scale deterioration of the earth's environment is now a big concern for all of us, and the situation has forced us to realize the urgent need for an adequate understanding of the dynamics of the earth. In this context, the significance of the ocean, covering about 70% of the earth's surface, and, in particular, that of the deep sea occupying about 88% of all oceans, is great. Despite their importance, the ocean and the deep sea largely remain a mystery as yet uncovered by man.

This session will discuss how we can achieve a better understanding of this mysterious area. I would like to have comments from each panelist about his interests in the deep sea. I hope we can address those topics shown on the transparency.

The first point to be discussed is how we can proceed on international cooperation in studies, of deep sea tectonics, deep sea biology, and material flux in important oceans. The Pacific Ocean is very important as a topic for future cooperation, because one of the most complicated plate motions is found there. And I hope we can discuss how we can cooperate in the next decade in research in the Pacific Ocean. This is the second point. The third is on measures we can take to promote workshops and other forms of interinstitutional cooperation. I hope we can have discussions on those points.

First, I would like to ask Dr. Dorman to make comments on matters related to deep-sea survey and research. Now Dr. Dorman.

CRAIG E. DORMAN: This morning I and several of the other speakers displayed for you much of the technology which we, in the past, have used for deep submergence with manned vehicles. This has dominated our science in the deep sea and has been going on now for more than a quarter of a century. Although much of our future work will continue to be with these manned submersibles, you have also heard earlier today about many of the new technologies which we will be bringing in order to enhance this type of research.

In my talk, however, rather than discuss those additional technologies, I thought I would show you very briefly the locations where the nations with deep submergence vessels have conducted their operations for the last 25 years as marked by these dots. The dots unfortunately do not show you the number of dives, and let me caveat my remarks by noting these are only the dives by the deepest of our vessels, by *Nautile*, by *Alvin*, by *Mirs* and by *Shinkai*. Some of the other submersibles, particularly those that are capable of diving only in the shallower coastal regions are not represented. But this does represent a general composite of all of the dives that we are aware have been conducted to date.

In spite of the fact that there are no numbers up there again, there have been roughly 3,000 dives made so far. The majority, as you can see, are fairly localized in terms of the areas where those dives have occurred. The vast majority have been along the major international, interocean ridge systems. The Mid-Atlantic Ridge, the East Pacific Rise, and areas off the west coast of the United States, where there are extremely interesting features. There are the straits of Juan de Fuca and further south there are some of the hot vent areas. Certainly there have been a very large number of dives in the Japanese waters of which you heard some details this morning.

As you will see, however, there are vast areas of the ocean in which we have not been, many places which we have not explored. Therefore in spite of the many findings and very exciting biological and geological features which we have uncovered in the past 25 years, we very strongly believe that there are many more extremely remarkable things yet to be found.

What I would like to do, therefore, is to show you two other viewgraphs. Both of these are for proposed expeditions. Whether or not these will actually occur I cannot say, but I offer them to you as examples of the type of thinking in the scientific community today as to the sorts and types of areas that we would like to visit and the rationale for those visits.

First, this viewgraph displays, in an overlay on the first, a proposed ship track for the Soviet vessel *Keldysh* and her Mir submersibles that we have been discussing with Professor Yastrebov's organization for collaborative dives during 1992 and 1993. The objectives of these dives would be twofold: first, to advance our knowledge in areas where we have previously been by detailed scientific studies in locales, particularly around hydrothermal vents in cold seeps that have already been studied by submersible. More importantly, however, we are proposing to go to many different sites, and first let me describe to you what the themes of our research would be. Then I will tell you a little about the major sites.

First, we are interested in exploring hydrothermal vents in cold seeps as yet unvisited. We expect to find new vent types, fluid chemistry, microbiology, and animal communities never before seen. We intend to explore large, isolated seamounts and island flanks in remote areas of the oceans, to study the benthic animal communities, cobalt crusts, lava flows in midwater, fish, squid, and smaller organisms. We intend to explore the deep fault zones of which you have seen a brief example this morning. These reveal windows into the lower ocean crust in the upper mantle several kilometers below the normal ocean floor.

We are interested in exploring active continental margins, including the Australian continental collision zone, to find cold water seeps, associative microbiology animal communities, and settlement deformation. We are also interested in exploring areas of deep-water upwelling to examine organic sediments, associative microbiological processes in midwater fish, squid, and smaller organisms.

Some examples of the places we are considering going include Mohns Ridge in the Norwegian Sea, where we expect to find new and isolated hydrothermal vent communities, and we will also look at sedimentation around and beneath the Arctic Ocean, and at the Arctic Ocean, midwater life, including deep-water corals. You will note that there have been very few dives in this locale. We believe that there are extremely interesting features to be found in these most interesting northern waters.

We intend to visit the Azores to look at sperm whales and squid around the western islands, lava flow on the island flanks, and animal communities on the slopes of the seamount in the islands. In the southern mid-Atlantic, we would like to revisit some well-known sites of marvelous diversity. These are called the Tag and Snakepit hydrothermal vents. We also would like to look again at the Vema Transform Fault for deep sections of the ocean crust and mantle, and at lava spreading from segments near that transform.

Offshore Peru where, again, no dives have previously been conducted, we would like to look for sediments and organisms beneath the Peruvian upwelling zone, the cold seeps on the Peruvian margin, and at deep crustal sections in the Hess Deep. This is one of the few areas where we believe it may be possible to actually drill through the crust and for the first time to be able to obtain actual mantle samples.

Offshore southern Chile, we intend to examine the collision zone between a mid-ocean ridge in the trench, looking at the lavas, hydrothermal vents and vent communities, deformations, and sediments.

Other areas that we are interested in include the hydrothermal vents near Easter Island, the deep crustal sections in the Easter Island microplate, the seamounts in the Austral group, and the Woodlark or Manus Basins east of Papua New Guinea. We would examine continental spreading of a volcanic arc environment and hot springs in small ocean basins. Northwest of Australia in Indonesia, we are interested in deformation, the Timoral continental collision zone and cold seeps in this collision zone, and sediments.

In the Indian Ocean, where again no dives have yet been made, we believe it would be most profitable to visit the Indian Ocean triple junction, looking at deep crustal sections near the Indian Ocean triple junction and hydrothermal vents on the mid-Indian-Ocean ridge. There are seamounts near the Seychelles, where we would like to look at the animals, the cobalt crust, and the lava flows.

The Gulf of Aden in the Red Sea has extremely interesting dydrothermal vents and Red Sea brine pools, again which have never yet been visited by manned deep submergence. Finally, in the Mediterranean and Black seas we are interested in the cold seeps in deep-water corals in the Helena Trench, the hydrothermal vents in the Aegean Sea, and the Black Sea sediments in the water column.

As another example of the type of collaborative investigations that we are interested in, I offer you again, as one possibility for additional collaborative dives, an area that we have been discussing most recently with our colleagues from JAMSTEC. We are proposing a month-long series of dives on the mid-Atlantic ridge in areas that, while we have been there before near the Kane Fracture Zone and its interception with the Mid-Atlantic Ridge, we would now have an ability to dive deeper than has heretofore been possible with our Alvin, concentrating in this area on the western intersection of the Kane Fracture Zone in the Mid-Atlantic Ridge, looking at the large fault scarps in that area as windows to the deep crust, and attempting to obtain unaltered samples. Further north we are proposing to look at some very interesting nontransform offsets, of which there are many in this area, but again which have been poorly studied.

For those of you who are not terribly familiar with this business, I offer my apologies for the technical terms and for the discussion in some detail. But I thought it would at least be interesting for you to understand that although we have been at this business now for many, many years, there remain an extremely large number of spots that have not yet been explored. We believe it still remains extremely profitable, in a collaborative international mode, to continue to approach the bottom of the ocean with our manned submersibles.

NASU: Thank you very much, Dr. Dorman, for the remarks.

Now the next panelist is Dr. Laubier. I believe that Dr. Laubier will speak from a global point of view, but in the Pacific Ocean, Dr. Laubier has had a very specialized involvement. So I think that he is expecting to share with us details about the Pacific as well as other oceans. Now Dr. Laubier, please.

LUCIEN LAUBIER: Mr. Chairman, first of all I would like to extend my thanks and my congratulations to all colleagues of JAMSTEC for their 20th anniversary, and I hope they will have new success in the future.

My talk will focus on one of the major discoveries of the last 15 years in biological oceanography, the occurrence of peculiar animal communities based on bacterial chemosynthesis, which have been discussed in several preceding lectures in this morning's session and just now by Craig Dorman.

Some 30 years ago when I visited the Bay of Tokyo at the occasion of the first diving cruise of the French bathyscarph Archimedes, which reached the bottom of the Kuril Trench at 9,500 m in July 1962, our knowledge of the deep-sea fauna was limited to zoological and biogeographical considerations. We knew that marine life occurred at great depths, and we also recognized the reverse condition; due to the lack of a sufficient food supply.

The first slide on the left illustrates what I can call the paradigm of the deep-sea ecosystem at that time, which clearly shows the very low efficiency mechanisms of vertical transport of organic matter synthesized in the Euphotic zone and automatically used by the venting deep-sea fauna. From the early 1960s, Alvin, was launched in 1964 or 1963 I think although we had a new generation of deep-sea submersibles using new buoyancy materials, the situation remained the same as far as deep-sea ecosystems were concerned up to the 1970s.

At that time, geologists came in with the plate tectonics theory, which in a very schematic way indicated two critical areas to be studied: the accretion ridge, shown in orange, and the subduction zones, in some kind of green I would say. Between these two borders, we know that the oceanic crust is slowly covered with sediments and deepens progressively. The processes of the accretion ridge were the first studied, and the 1977 dives of the famous Alvin, near the Galápagos Islands, showed the extraordinarily dense communities of marine invertebrates living in the vicinity of warm water of this charge. It took 2 years before biologists entered the game, in 1979, with the first cruise devoted to life sciences.

This slide shows, you know them I suppose, the two submersibles used by French teams, Cyana on the left, which was built to go at — it is still there, by the way — to go to 3,000 m, and the Nautile which can go to 6,000 m.

New tools had to be developed for biological purposes. We had to develop those shuttles used to carry material from the deck of the ship to the bottom of the sea and back to the surface. We had to develop closing baskets, because all the material is very light and could be lost at the surface or on the way when coming back, and so we need baskets with a closing device. We also had to develop a temperature recorder that is placed on the bottom and has a probe connected with a recorder, with both cables put on the bottom by submersibles, and sediment traps to record the influence of the chemosynthetically synthesized organic matter. These are essential tools for our work, including, of course, microphotography and special *fluid* samplers, and this shows you our way of modifying our submersibles. This is the *Alvin* with all front equipment for micro-photography, and this is a special set of six bottles for keeping pressure that are used to study the eutrophic thermophilic bacteria as handled by the *Nautile* in this slide.

The major point in this early period was achieved in 1979 with the discovery on the East Pacific rise of the black smokers at 350°C. This is the *Alvin*, and this is *Cyana*, in the same area on the East Pacific rise.

The first point that interested the oceanographers was the very high density and biomasses of the hydrothermal vent communities. It was immediately demonstrated as well on rocky substrates, at 13°N on the EPR, East Pacific Rise, as well as on sedimentary substrates. This is the Gulf of California, and you can see the Guaymas Basin with the same groups of animal. The living sides are very limited in space. They are also very limited in time. The total surface varies between a few tenths or a few hundredths of a square meter, and the active life of a given chimney does not exceed 1 century.

On the left side you can see the scales. The bar is 1 m, and you imagine how we need manned submersibles to study that sort of very small scale spot. On the right side there is a graveyard of shells which frequently occur and recall previous living *forces*. We know now that the physiological study of the giant clam and the large live worms demonstrate that they receive their food through internal symbiotic bacteria while others, such as the worm Alvinellidae, are using this old organic matter produced by antibiotic bacteria.

Another important discovery was made in 1984 on the Oregon coast and then by Nautile 1 year later in the Japanese Trench. These are the oases of large bivalves associated with symbiotic bacteria found between 3,000 m and 5,900 m at that time. I heard yesterday that a new Japanese report puts the ultimate depth of *Calytogena* at 6,350 m.

Vise versa subduction causes cold seep. The hydrothermal communities, we know now, are more diverse than cold seep communities. Here I show you just two examples, one in the Pacific, in the western Pacific, studied by the Japanese-French program STARMER as we heard this morning from Dr. Uchida. This one spot we have called the "White Lady" because of its general color, and on the right there is an interesting fact that the functions of chemo-synthesis symbiosis by the large worm are replaced by two large gastropods: in the center is *Alviniconcha* and outside IFREMER, living at the same depth as the tube worm.

The next slide shows you another example, and the last example, in the Lau Basin, very near the North Fiji Basin, at a depth of 2,000 m. It shows spectacular colonies of a living fossil, the cirriped, and also the very interesting shape of the chimneys. This is a place where we have recorded the highest temperature, about 400°C. The deep-sea cause relies on chemosynthesis achieved by symbiotic and free-living bacteria. Many important questions remain to be solved, such as the propagation mechanisms and evolution processes. Also, the biotechnological potential of *eutrophic* bacteria seems important.

As shown by previous speakers, potentially interesting sites are numerous all over the world. There is a growing scientific interest in the deepest parts of the trench, from 6 to nearly 11 km in depth. I hope international cooperation will enable us to explore the feasibility of new deep-sea areas using manned submersibles for such ultimate depth, as well as better coordination of nations' efforts.

Remotely operated vehicles and robots are very welcome in ultra deep sea, but in my mind I still think human eyes and brains behind a porthole are an essential tool for future deep-sea exploration.

NASU: Thank you very much, Dr. Laubier, for the comments.

Now the next panelist, Dr. Yastrebov, who will talk from a physical and chemical point of view. We would like to hear his comments.

VYACHESLAV S. YASTREBOV: Thank you very much, Mr. Chairman. I would like to talk about some results of study in the near-bottom layer. Processes in the near-bottom layer are highly variable in time and space. The near-bottom layer is an open system to which mainly hard suspended material is continuously transported and from which the solvent dissolves and, more seldom, suspended matter is continuously created. Biochemical activity of the near-bottom layer is caused by the accumulation of large bodies of both living and nonliving organic matter and *endogenic* matter.

The chemical composition of the near-bottom is formed under the most essential influence of chemical exchange with bottom sediment and endogenic effects of the Earth's interior. The other bottom chemical interaction mechanism is determined significantly by diffusive exchange increased with currents, and the exchange is the result of differences in the concentrations of both elements in the liquid phase of the sediment and the near-bottom water.

Three special models of the near-bottom layer are known:

1. The water above the sediment is well mixed, and concentration of its properties is evident.
2. Immediately near the bottom a layer is distinguished with laminar motion and high concentrations of matter resulting from its molecular diffusion from bottom sediment.
3. The upper layer of sediment is bioturbated or reworked with the bottom current. Chemical reactions in it are highly accelerated. The contiguous water and bottom layers interact intensively, and the upper active layer of sediment becomes repeatedly turbid, and undergoes physico-chemical and biological transformations. This model has been supported experimentally.

The main aspects of the near-bottom layer dynamics and problems of a highly energetic layer are dealt with in brief in the present paper. Discovery of highenergy boundary layers was one of the most important events in the oceanology of the last 2 decades. More exactly:

1. On the large areas of world ocean, the benthic layers of 50–150 m, in height were discovered, which were characterized by elevated, sometimes essentially velocity, high turbulence energy and strong internal mixing.
2. Within these layers the distribution of salinity, temperature, and concentration of sediment particles are uniform. In this picture, you can see the distribution of temperature, salinities and transmission, in the experimental materials.
3. At the upper boundaries of these layers a very sharp appears. This reduces exchange intensity practically to zero level. Due to this ceiling the water masses of benthic layers move without mixing over distances of planetary scales.

We understood the mechanism behind this ceiling and we formulated the mathematics of lower quasi-homogeneous oceanic layers, as lower pycnocline, including physico-chemical and biochemical processes. This model explains all features observed in benthic boundary layers.

I will explain this very point in more detail. We showed that flow stratification by suspended sediment plays a governing role in the ceiling. This gives an essentially new sink of turbulent energy-work against the buoyancy force, and leads to a relation for turbulent energy. In this picture, you can see these characteristics and sediment units, a very small sediment unit and a full sediment unit.

The basic result is that K_0 is close to a constant and suddenly starts to grow to the value 1 at the boundary of the benthic layer, the exchange coefficient A vanishes, and a ceiling occurs. I stress that we do not assume much variation of the exchange coefficient. We introduced a new physical factor, stratification by suspended particles which are available everywhere, and we obtained such behavior of the exchange coefficient.

I will show you the results of calculations based on the mathematical model proposed, for realistic values of parameters, the sediment concentration distributions, and exchange coefficient distributions over the height of the layer. The case was considered when the sediment was suddenly supplied at the bottom. We see clearly the formation of characteristic stepwise concentration distribution, vanishing at the turbulent exchange coefficient at the upper boundary of the layer. Neglecting dynamic effect of particles we obtain an absolutely different distribution of concentrations. You can see the total characteristics in this picture, no sediment, no step, no ceiling. Thank you very much.

NASU: Thank you very much.

We are going to hear from Professor Kobayashi. He will mainly discuss the activities in and around Japan. He is quite active in the plate survey, and perhaps we will be able to hear something on this aspect as well. Professor Kobayashi, please.

KAZUO KOBAYASHI: Thank you. As introduced, I have been mainly working on the Northwestern Pacific Ocean floor, around the Japanese islands. I would also like to refer to our future activities as well.

As indicated on this slide and as is used on the cover of the JAMSTEC brochure, this is a well-known topographic chart of the Japan Trench. It goes all the way to the Mariana Trench. This is the Nankai Trough and the Ryuku Trench, and beyond there will be the Philippine Trench. So there are deep trenches where the ocean floor of the Pacific and the Philippine Sea are subducting at the rate of 2–8 cm/year. There is a lot of tectonic activity going on here, and this brings about a lot of disasters, serving as the source of volcanic activities or earthquakes in Japan.

So what is necessary is to thoroughly investigate and survey these sources and activities. In order to do this, research vessels, of course, will be necessary. This is a drilling ship, which drills the hole on the ocean floor and seabeds to investigate the situation and phenomenon of the seabeds. This ship has been used for some 20 years under international cooperation, in areas including the sea around the Japanese islands. Here, deep-sea submersibles are also used. So three faeilities are used for the deep sea research. From the surface of the ocean, to drilling, to looking at the inside of the seabeds and also diving with a submersible to actually take measurements on the seabeds. These methodologies have been used for the survey and, as a result, as far as the trenches around Japan are concerned, a lot of information has been obtained, and active faults have been discovered.

Perhaps these activities have been going on for the past several years, tens of years maybe, sometimes as many as hundreds of years from a geological viewpoint. At least compared to the age of the Earth, these phenomena, which can be observed underwater, are relatively new. In the case of the Nankai Trough, we are working together with French scientists. There was the "Kaiko" Project and discussions are ongoing to further extend this project. As for drilling, DSDP has been carried out, and also at the Nankai Trough, and tremendous achievements have been realized.

Around the Japanese islands, research targets are not just in the trenches, but there are also a number of areas with interesting topographic features. It seems that all the special features around the world are concentrated in these areas, one of which is on the oceanward slope of the trench, to the west of this arca. There, fresh cracks were discovered by the Shinkai 2000 submersible.

Another example is on the back-arc area of the Ryukyu Trench. In this area the sea floor is starting to crack or rift, so it is a similar situation to an egg hatching. I think you saw the photograph of the black smoker, which exists in the rifting area. Shinkai 2000 observed this clearly. High-temperature magma rises to near the sea bottom, the water is heated with this magma, spurts into the water, and a massive amount of sulfur is spurted as hot water venting. Sometimes mineral ore deposits are formed around hydrothermal venting chimneys. Also, there seems to be another area of this type of phenomena to the west of the Ogasawara Trench. Such observations have also been reported by drilling investigations.

There is another active phenomenon, again a new discovery made around the Okushiri Ridge. This is a possible trench which seems to start and has been subducting for some 1.8 million years. This idea has been proposed through a study on topography at first and Shinkai 2000 confirmed some evidence. Later, drilling was done to determine the age. We have obtained very significant results by use of this three-stage approach.

In addition, around Japan there are older and now inactive phenomena observed which used to be active in the past, for instance, in most parts of the Sea of Japan. Two years ago we did ODP drilling and the results revealed that about 15 million years ago this area rifted open. In olden days the Japanese archipelago used to be part of the continent, but then it started to drift away, split away, possibly. The age of the phenomenon was able to be determined by using drilling data and the process also was uncovered. A similar phenomenon was confirmed in the Shikoku Basin and the West Philippine Basin. Also further efforts shall be made along the same line so that we can discover similar phenomena around Japan. However, it is not enough just to observe the area around Japan. We also need to compare such data with international data.

There is a measurement from the surface using acoustic methods that enables us to do this, submersible observation is another form, and then there is drilling. At the Nankai Trough, we did drilling, and the hole is shown in cross-section. Toward the bottom of the trough, you can see the mud, and there is the trough or rather the fault, and somewhere deeper, an earthquake was also occurring. So in such a hole, what we plan to do in the future is to install some instruments and measure the change in temperature and do electrical measurements on a long-term basis, which will become a necessity. This will be an underwater station, as it were.

We used to have stations on land, but now by drilling holes we will want to install these instruments and equipment in the hole, and in order to recover such equipment, submersibles will be very helpful. So again the three different types of approach will be used in combination in the future. This will be done not just around Japan but all over the world. In order to understand the interior of the Earth also, wide distribution of such observation equipment and facilities will be necessary.

Further, in another important study around Japan, in the Sea of Japan and also in the older part of the Pacific, sediments will be sampled or recovered with ocean drilling, and by analyzing such samples the history of the Earth since the seabed was formed some 180 million years ago will be inrestigated. In the case of the Sea of Japan, events maybe since 10 or 20 million years ago will be uncovered. In particular, we will be able to get all of the information as to what has been happening since the Ice Age, Thank you.

NASU: Thank you very much.

Now I would like to ask Professor Karube to talk about the area of biotechnology.

ISAO KARUBE: Thank you very much. My name is Karube.

Dr. Laubier mentioned there are a number of organisms in the water, and we are trying to use the organisms, as an engineering application in bioelectronics, as the area that I specialize in. So let us look at the topic of oceanography from that point of view.

I am working at the Research Center for Advanced Science and Technology, University of Tokyo, which is a 4-year-old center for the study and development of advanced technology. Initially, our main interest was to conduct marine biotechnology studies, and Toyo Suisan Fishery donated us ¥200 million whereupon we opened up this research division. From Dr. Baker's institute, there are three visiting researchers. We have an associate professor and an assistant professor.

We are interested in fixation of CO_2 in relation to global climatic change, and we are also interested in substances produced by marine life that block ultraviolet light. We have also established an academic society for ocean biotechnology. Dr. Miyaji is the president of the Japan Society for Marine Biology, a society which organizes study groups to promote the study of marine biology.

The Ministry of International Trade and Industry established the Marine Biology Institute. Some 24 major companies are part of this research institute, which conducts basic study, and I myself am involved. Bioelectronics is my main area, and I would like to show you some slides.

This is the Marine Biology Research Institute, one in Shimizu, the other one in the city of Kamaishi. This institute looks into a number of substances taken from marine life, such as the use of microorganisms to clean up the water after a petroleum spill that may contaminate the ocean. There are also a number of other studies. We have not investigated deep-sea organisms, but we hope to conduct such studies in our future research projects.

This is the ship that used to belong to the University of Tokyo and we completely refurbished the facilities of the ship for marine biological study. This vessel collects a number of organisms from oceans of the world to be used as materials for our study of marine biology. Next year we are planning a collaborative study with France, and last year, together with an Australian counterpart, we conducted joint research. We hope to expand the scope of our joint research work with a number of other nations.

As many of the previous speakers have rightly pointed out earlier, aquatron and other equipment and instruments are necessary in conducting our studies. Japanese advanced technology may be utilized and incorporated in these facilities. We collect a number of organisms from the deep sea and culture them at very high pressure. Some of the facilities are still under development, but I think we will be able to simulate the conditions of the deep sea using these high-tech facilities in the future.

This is just an example of possible study areas. We are looking at a mechanism so that a lot of organisms will not stick to the bottom of the vessels, for instance. This happens to be a project of MITI, and biologically active substances and farming of fish cannot be part of the projects. Private companies have contributed their own money to conduct studies on biologically active substances and marine culture. This is a rather bold plan.

The global environment is a grave concern for once, and carbon dioxide may be fixed and we may be able to turn it into calcium carbonate. In that way, we will be able to take advantage of CO_2 rather than seeing it as a substance with adverse effects. Fixation of CO_2 and conversion of CO_2 into calcium carbonate are quite interesting, and we are trying to identify organisms that may contribute to this research area.

The other interest is in functions of marine life which will be applied to electronics. This is what is called a biosensor, and this is attracting a lot of attention around the world. We are using biological elements and combining this with electronics. We can make measurements of an organic substances. For instance, in the deep sea organic substances are produced and if we install a biosensor in deep water, we can make measurements of the organic substances produced in the deep sea. This kind of biosensor may be used in the field of oceanography.

The deep sea is a very difficult and harsh, hard environment. A lot of these organisms live under conditions of high temperature and high pressure. We can take some chemical components from these organisms, and these components are probably more stable than the substances obtained from organisms living on land, a property which may be exploited in some way or another. So, in that sense, I think deep water can become a very interesting field for our study. In fact it is already being developed, with some 80 companies involved the development of biosensors.

In the field of bioelectronics, transistors and microelectrodes are used as transducers. This is an area which Japan is very good at. Japan is very strong in electronics, so by using advanced electronics in combination with the functions of marine biology, there are a number of potentials that we may be able to exploit in the future.

This is an example of a micromachine, where semiconductor processing technology is used, with the sensor-actuator combined in one unit. This is a very compact unit, $1-10\,\mu m$ in size. Using this, we can measure chemical substances. This is a turntable. You do not need to use large-scale liquid chromatography; you can use this micromachine to make the same kind of measurements. The development has just started, but we can probably install this small piece of equipment in deep water and pick up signals from these instruments. Using a remote-sensing device, I think we will be able to understand the kind of chemical substances present, organic and inorganic, and how they are distributed in the deep water. I think that can be a possibility in the future using micromachines.

This is yet another example of the function of marine organisms. There are some very interesting molecules, and it may be possible to incorporate that function into a computer. So we may be able to build a biocomputer using elements like that of marine organisms, sea hare being an example here. Using sea hare, combined with advanced technology, a sea hare's information-processing function can be elucidated. We can combine that information-processing function with a computer to make a neurocomputer. Squid are another example. This information-processing mechanism of a number of organisms can be used to develop a neurocomputer or some computer which is completely new in concept.

We are extremely interested in the deep-sea survey, marine biotechnology in some day will concentrate its efforts on deep marine biology. Thank you very much for your attention.

NASU: Thank you very much.

Now our last panelist is Dr. Nakato, and I think that he will speak from the point of view of observation in the ocean mostly. Dr. Nakato, please.

HIROYUKI NAKATO: Thank you for the introduction. I am Nakato. I am from JAMSTEC, and my expertise is in the areas of deep-sea research and exploration technology. So I would like to consider the prospects as well as the desirable directions of deep-sea research and exploration technology, including my personal views.

I would like to show you some slides. The slide here shows the submersibles that we at JAMSTEC own and operate. Shown here are *Shinkai 2,000* and *Shinkai 6,500*. Not shown here is the 3,000-m-deep remotely operated vehicle (ROV) named *Dolphin 3K*. As Dr. Uchida said this morning, we are currently in the process of developing an ROV with the capability of diving down to 11,000 m. That means that will be capable of diving in any deep ocean anywhere in the world. It is expected to be completed 2 years from now. So using these facilities, we would like to elucidate the state as well as what is happening in the deep-ocean beds, where human beings have never been. But for these explorations, the regrettable factor is that there are too few manned submersibles available today.

At present, the most accurate and sharpest observation capability exists in the human eye. Therefore, we believe that sooner or later the construction of a manned submersible with the capability of an 11,000-m dive will become a reality. Dr. Laubier stressed the need for the next-generation submersible, a manned submersible, to be developed, and I think that we see increasing international consensus or, shall I say, international public opinion stressing the need for developing the next generation of manned submersibles.

In the area of unmanned submersibles, or ROVs, we have cable-free ROVs available today, those that will be available for practical applications in the near future, and also autonomous-type ROVs with a built-in computer that will be applied to practical operations very soon.

In the deep-ocean beds, many changes are taking place, such as changes in deep-sea life and in topography at the seabed. They are taking place over an extremely long period of time, so the duration of time that can be covered by manned submersibles and unmanned submersibles is too short to trace the changes at the seabed. What we need today, very much, is the development of long-term continuous observation systems for long-term observation at the seabed, which Dr. Kobayashi also referred to.

At JAMSTEC, what you see here on the slide is our ongoing efforts to develop a long-term continuous observation station at JAMSTEC. This is the long-term continuous observation station. It is a conceptual image. The size will be about 3 m and 1.5–2 tons will be the maximum, ultimate weight. Housed in the station will be a tiltmeter, seismometer, ocean-water temperature monitor, and other instrumentation on board the station. The ultimate development expenditure will be about ¥100 million, a very expensive project.

This slide shows the technological challenges for completing such a long-term, continuous observation station. They are as follows: In the first place, the station must be capable of making long-term and continuous records, long-term record making. Secondly, the station must have the monitoring capability on a real-time basis, or near real-time monitoring capability, as well as a near real-time or real-time transmission capability, to be able to transmit the observed data. Thirdly, the station must have an adequate power source to support the long-term observation. The sensors and other instruments to be utilized must be hyperbaric; they must be able to withstand very high pressure levels.

In the ocean beds which are close to the land, these facilities can be connected by cables. Therefore, these four problems that I have outlined in the previous slide can be solved by cables. But when it comes to the offshore areas, including the trenches which are far away from land, cables cannot be used. Therefore, in the case of such stations for long-term offshore observation, what we have to achieve are large-capacity memory and data transmission systems with support by satellite and voice and adequate power source development. These stations must be deployed and maintained or repaired or supplied with necessary materials, and these will be done by existing, manned submersibles.

What we need is not just long-term observation, but also to expand the coverage of the observation areas. In other words, we have to develop wide area observation. This requires us to make a network of long-term observation stations. At present, JAMSTEC in cooperation with Woods Hole, is developing A-B-E or ABE, which stands for autonomous benthic explorer. Once completed, ABE will make observations in areas in the vicinity of such a station so we look forward to this great role.

The ocean bed is the habitat of many microorganisms that we do not know on land. Therefore if we can reproduce these fauna on land by culturing and incubating them we will be able to understand much more about what life exists on the seabed. So at JAMSTEC we are in the process of developing an incubator to reproduce those seabed microorganisms on earth, to expand high pressure, low temperature, and high humidity. On the seabed, many other natural phenomena are taking place, such as volcanic activity and life communities. They originate in the sub-bed, which is more difficult to observe than the deep seabed. In this area, a drilling vessel will play an extremely important role for sampling the subseabed material and to put in the sensors and instruments in the drilling hole to know what's happening beneath the seabed. The United States has already developed and is operating such a deep drilling vessel and has achieved many accomplishments.

In addition to the existing technology of deep drilling, perhaps riser technology needs to be developed for more depth drilling and observation. Therefore, we at JAMSTEC believe that we must develop such a deep drilling vessel or a driller, and our objective is to realize it by the end of this century. At present we are carrying out a feasibility study on constructing such a driller. The slide here shows the concept of the deep drilling vessel we would like to develop. The size will be 15,000 tons, the maximum depth of its drilling 7,000 m, and it will be able to drill another 3,000 m in the subbed further below the 7,000-m level.

The blue and black areas indicate the area to be drilled, the deep sea and sub-bed. The riser tube will be installed here. Drilling will be done in the seabed, and natural gas may erupt from this sub-bed, so there must be a mechanism to prevent the rush of natural gas or oil from the sub-bed. These technology elements need to be developed before we can complete this system.

In the area of developing deep ocean bed drilling vessels, Russia is now constructing one and France also has plans to develop a drilling vessel for applications on the ocean bed. However, one single deep drilling vessel cannot achieve all purposes. A multipurpose vessel will be very difficult to accomplish, so more than one deep drilling vessel needs to be applied for many different purposes. Here I see the potential of international cooperation with the participation of many countries to capitalize on and complement each other's capability.

The ocean is a vast area. There must be the critical mass in the first place in facilities such as manned submersibles or ROVs. There must be a sufficient number of vehicles, and there must be efficient utilization and operation by more than one country through close cooperation. The budget for building a 10,000-m manned submersible is very expensive, so therefore a deep drilling vessel can be constructed as an international joint endeavor. Observation of plates and ridges and trenches must cover a broad area of the seabed, and to turn it from a concept to a reality here again I believe that international cooperation or joint endeavors are essential. We are very happy to have in the symposium the representatives of the leading oceanographic institutes of the world, so I believe that we have a potential for collaboration.

NASU: Thank you very much. We have heard comments from the panelists. There was new information, new and valuable information. I would like to thank you for providing us with very instructive information.

Let me summarize the comments made by the panelists. There is the seafloor and on top of that there is water, which is a liquid phase. Below the seafloor, there are rocks and sediments. A lot of comments were made on those layers about the deep seafloor.

What sort of methods can we use to do surveys on that near seabed layer? Submersibles, either manned or unmanned, can play important roles in the surveys. At the same time, deep-sea drilling vessels can be constructed in order to do surveys of the subseabed. They will be quite important as well. We can also use sonar to do surveys of the geological structure. As well as these physical measures, there are chemical measures that we can resort to, and by utilizing these physical and chemical means we can study the physical properties of the liquid, as well as the particles in the liquid phase.

We can also study the sediments and rocks. We can collect samples of sediments and rocks or sometimes we can leave the apparatus for a longer period of time in order to study temporal, or time series, changes which take place under the seabeds. I think some people pointed out the importance of that sort of longer term study.

Anyway, at the present stage, observation stations or observation points are lacking. From various perspectives, I think, the panelists were emphasizing that at present there is a shortage of observation points.

So having listened to the comments made by the panelists, and based on my past experience, I would like to share with you my impressions. It was in 1959 that the Japanese Expedition of the Deep Sea was started using the *Ryofu Maru*, which was the former Research Vessel belonging to the Meteorological Agency. It was a fully-fledged deep-sea survey in the post-Second World War period in Japan. At that time, in the bed of the Japanese Trench, which is 6,000–7,000 m deep, there were a lot of sea cucumbers, though the shape or the shapes were unusual. There is a Professor Masuohi Horikoshi on the floor today. Now I recall that I worked with Professor Horikoshi to sample the cucumbers.

In 1967, in the Ocean Research Institute of the University of Tokyo, the former *Hakuho Maru*, which is still used in some research institutes, went on an overseas expedition for the first time to Hawaii. At that time, Professor Yasuji Katsuki, Tokyo Dental and Medical University and ourselves visited Professor Georg von Békésy in the East-West Center of the University of Hawaii. Professor Békésy is originally from Eastern Europe, and he was granted a Nobel award for auditory sense. When I told him that we had found many sea cucumbers in the deep spots of the Japanese Trench, Professor Békésy said it is quite scary. I have a strong impression of that. He said it is quite scary. Scary.

At that time the professor might immediately have had an idea that there existed the possibility of different kinds of biological processes under enomous amounts of pressure and in darkness, if compared with that under the atmospheric pressure on land.

In the deepsea, around black smokers and white smokers, there are a lot of life forms which are completely different from those on land and in shallow seas. They have different physiological properties, but marine animals have totally different physiological properties from those organisms living on land and shallow sea. So we had found something quite different, and Professor Békésy in 1967 must have foreseen that difference, and that is why he must have said it is scary.

Manned submersibles can go as deep at 6,000 m, and a lot of good research results have been obtained using them. However, I think in the future there will be manned submersibles which can go as deep as 11,000 m. The new submersibles to be constructed will be different from the Trieste or Archimedes in that they will actually have arms and will be able to grab things from the seabed. Of course, as Professor Laubier said, Archimedes went as deep as 8,500 m in the Kuril Trench, and the sister vessel, called Trieste, which was transferred to the American navy, once went as deep as 11,000 m in the Mariana Trench.

I would like to tell you one thing which not many of you know. The captain at that time was Mr. Don Walsh, a famous officer of the American navy, and the scientist on board was Dr. Piccard, the son of the famous Auguste Piccard. The Trieste went more than 10,000 m deep and then landed on the seabed. That was the only time, it was the only time the *Trieste* could do it. Let me tell you the reason why *Trieste* was successful in doing so. There was a leakage of water inside the *Trieste* at shallow depths. Of course, there was a lot of concern about the leakage inside the submersible and when it came to the surface an instruction was sent from Washington not to submerge again because it is dangerous to go below water when there is a leakage. So that is why it did not go underwater again. There is higher pressure if you submerge, so if there is any sort of slit through which water seeps in, the pressure will cause the slit to close. The *Trieste* could therefore submerge as low as 10,000–11,000 m without any problem. That is the interpretation by Captain Walsh. In the future when submersibles with arms are constructed, we will certainly be able to go below 10,000 m and I think that will bring about new research results and survey results.

While listening to the comments made by the panelists, I thought of various events which took place in my long-term experience and experiments in the last 3 decades. I was involved in the formulation of this trench project and the Glomar Challenger. There was a project involving international cooperation, called IPOD, for this Glomar Challenger, and I was involved in that.

I am also here as a kind of representative of the IOC of UNESCO. Because I have served as a Japanese representative of the IOC in the past 9 years and because of my close relationship with the IOC I was asked to attend here. Of course, Dr. Kitazawa is here as the formal representative of the IOC.

Through my experiences of international cooperation, there is a famous organization called ICSU, and the Science Council of Japan is the contact point in Japan for this cooperative project. Of course, academic wisdom is important but budget is another important factor in this sort of international project. We can ask for funds from the private sector. But at this time we require budgets or funds from the government with funds from the public sector also playing an important role.

On the other hand, in the IOC, which is not a nongovernmental organization in this sort of formal intergovernmental forum, all country representatives get together to discuss various topics, and each representative goes back to his own country and the topics are discussed again. Then there is the implementation stage. But looking at this process, in the case of France, IFREMER is a very strong core institution. In the case of the United States, Woods Hole Oceanographic Institution and Scripps Institution of Oceanography are playing leading roles. In case of Russia, the Shirshov Institute of Oceanography must go as the major institution. In Japan, many universities, many institutions, and agencies of fisheries and other institutions are involved in research, but JAMSTEC and the University of Tokyo's Ocean Research Institute are the strong and powerful institutions.

Consultation and exchange of opinions among these leading research institutes has taken place in the past and the exchange has borne many fruits. With today's symposium, even if it is 1 day or 3 days, after the symposium is finished we just say good-bye, after acquiring a lot of knowledge and after exchanging information we go home to our own work. But I think that is not enough. Certainly, we can do more and we can obtain more from this sort of experience.

Why not take advantage of this opportunity in order to promote cooperation on a permanent basis among leading institutions in the world? For example, why not form a workshop on a permanent basis? Sometimes, they could serve as ad-hoc workshops. In other words, we could discuss certain topics on an ad-hoc basis in workshops, or workshops could be used for longer term objectives. I think that in that way this sort of symposium will bear bigger fruits. I do not like to sound presumptuous, but since I am serving as the chairman of this panel discussion session I would like to suggest the idea of forming or organizing some sort of workshop in the future. I think that will help to promote oceanographic research and development a lot further in the future.

I have one suggestion — interinstitutional or interagency cooperation. A very strong cooperative relationship needs to be implemented, while fully respecting each participant in the agency or organization. What is most important is good communication, and this will enable smooth communication. What will be the ideal way to go about such interagency cooperation? Perhaps we could choose this topic as something for us to discuss together here, so that we can improve our communication. Perhaps Dr. Dorman, could you start with a short comment please?

DORMAN: Yes, thank you. I find your idea intriguing, I must admit. We have I think, in a very small way, the elements in place already for hopefully doing more along the lines that you have suggested. Most of our relationships, of course, in the past have been on a bilateral basis, and I think many of us have already, and my friend Dr. Laubier I am sure will say more about this, obtained agreements that could lead themselves to expansion to improve more of our operational people.

One particular mechanism that comes to mind is where all of our institutions have previously met as a group, under the agreement that Woods Hole has with JAMSTEC. One of the topics of concern has been the safety of submersible diving, and we have held symposiums on this topic. The most recent, I believe, was at Woods Hole about a year or so ago. We have discussed the possibility of broadening the scope of this to include other submersible types of issues. This may be a possibility, simply because we have done it before and it is a schooled way to perhaps continue, expand, and get the technical operational people together.

I would urge that if we do this, and I do think it is a good idea, that where it may profit us to start our discussions is in a nongovernmental sense, if you would, looking at the issues of the technology and operations from the perspectives of the operators who are the most familiar with the vehicles and equipment. Perhaps we can make most progress early in the stages of such discussions if we leave politics and institutional issues aside and deal primarily with the issues of those who go to sea in and with these vehicles. If we can capture some of their dreams and views and concerns, that I believe would perhaps serve as a sound basis for us then to look at our feasibility and our mechanisms.

My experience has been that if we start the other way, we very soon and very early bog down because there is a tendency to perhaps form plans that are not based solidly on the reality of the technical and operational characteristics and capabilities. So I would argue that we start by dealing with known issues. Safety is a good example. That is a primary concern that we have all had, and one that we have been able to talk about irrespective of our national interests. Those of us who are in those vehicles have a very firm interest in it, and I think perhaps an expansion along those lines might be a good way to begin.

NASU: Thank you.

LAUBIER: Thank you. I have one comment following on well from the comment by Craig Dorman. If I look in the past as to what has been done in terms of international cooperation in the deep sea, we have had discussions with the United States about safety problems for deep-sea submarines.

If I look at science programs, I see that we have just finished with Japan a program of hydrothermal processes in the Back-arc Basin of the North Fiji Basin. We have in mind to do something more now, still in the Pacific. At the same time, we heard about discussions between the same Japanese people from JAMSTEC and our Woods Hole colleagues.

So I think if we consider that we have also in France some kind of bilateral cooperation with the United States on the mid-Atlantic ridge, we have the three sides of a triangle. I really believe myself that if we could unify our efforts, not only by two but maybe by three, and open widely the use of those tools to other nations who have not acquired that kind of technology, we can make progress.

The last idea I would give to the floor is that, as you understood I am very keen about the manned submersibles, but I realize that this is not the only tool we need. We need robots, we need permanent observatories, which have been developed by Director Nakato. I think they are very important, too. But in my mind it is more or less time to see what could be a further related study of new material applied to submarines, manned submarines, just as a feasibility study. We are making feasibility studies for a riser, a drilling ship. Why not do the same for a new type of manned submarine? See if the limit is 8,000 m, 10,000 m with the new materials we have in hand. This is something at least three or four countries could do without a large amount of money, and that could help to make a decision in a few years.

NASU: Dr. Yastrebov, have you a comment please?

YASTREBOV: For my opinion, I think that it is necessary to make two things. First, we must organize a joint program, on investigating the ocean. Second, we can make a formal agreement on world cooperation, about international cooperation. These two things are fundamental for our development in the future. Thank you.

NASU: Dr. Kobayashi, please.

KOBAYASHI: As was already pointed out, when it comes to the ocean, in particular deep ocean R&D, it requires a variety of tools and equipment. That was already the case in the past and even more will be required in the future, which means that a single country alone will not be able to go on working on its own. So inevitably, cooperative work internationally will be necessary. In order to achieve a form of workshop I think we should take every opportunity to realize this, and I think we should have further exchange of opinions and views. I think that will lead to realization of international cooperation in the true sense.

However, if we try to include everything in our approach, to try to do everything or too much, we might get bogged down. So from time to time, the most feasible topics should be selected for incorporation. I think that will be a key to success on a long-term basis.

NASU: Thank you. Dr. Karube, please.

KARUBE: I would like to think in terms of international cooperation in the area of marine biotechnology. In Japan, there are American or Australian researchers, and also starting in the early part of next year, we will be having American as well as German researchers. So between universities, there is quite a lot of active international exchange going on.

However, when it comes to joint research started between private industry and the government, there are barriers or obstacles. For instance, Japan does not acknowledge 200 nautical miles whereas Australia and France do recognize this. So there is this problem related to commercial issues. That is an obstacle and we are having some rough going at the moment.

As far as academic joint cooperation is concerned, if we can pursue those lines successfully I think it will be a good way to start international exchange. What we want to see is this: for instance, using a deep-sea microbe to conduct a biotechnology project would be a very interesting.

Dr. Koki Horikoshi, at one of the institutions, started a new project using a variety of deep-sea microorganism samples to investigate their functions. This will be a basic research project, and then after that as a next step using microorganisms, he wants to try making different things under very stringent conditions. If the microorganisms can still survive in culture, to make use of them will be a very attractive idea. But to do this, we need a deep-sea submersible.

We have *Shinkai 6500* but, for instance, if we have someone from France asking for samples from the deep sea or if someone from the United States wants a sample, I think perhaps cooperative or joint research in the area of provision of samples will be one possibility. I think we can have a give-and-take relation where Japanese samples can be given in exchange for some other samples, and also a data bank or database. Perhaps we can come up with a unified or common database through activities of such a workshop. I think such efforts will be very important in the future, so that's my proposal.

NASU: We have solicited views from the panelists, and Dr. Dorman suggested, for instance, the safety of submersibles as a starting point for the workshop. We can also develop various other themes for the workshop. I would like to request Dr. Nakato the following: JAMSTEC organized the symposium and, of course, you are not in a position to decide which themes may become the subject for the workshop but I hope that Dr. Nakato and other staff at the JAMSTEC will take the initiative for such a workshop. I hope that Dr. Nakato is willing to take that initiative.

NAKATO: I am more than happy to take that initiative, as I stated earlier. The ocean is vast, and there are so many targets that we have to look into. International collaboration and cooperation is essential. Otherwise, there is no development in the study of ocean and utilization of ocean resources.

　　We have to go down to the essentials of what themes we would like to take up and at what speed we may like to continue on with the workshop projects. The chair suggested a possibility of the workshop, and if an agreement may be reached I am very happy. I am also very happy to act as a caretaker.

NASU: Thank you very much. There are strong oceanographic research institutions represented here, and it is important that such collaboration may be continued. The reason why I am emphasizing this point is that when we hear the word oceanography, there is no word like landography which corresponds to terrestrial study because, study fields on land have already been split into many specialiged areas. That means mankind's involvement with the ocean has often been rather limited. There are so many things that arc still unknown to us, and we have to uncover the unknowns. There are a lot of areas for us to study and look into. We have to make sure that our net is really tight so that everything is uncovered.

　　In December of last year I went to Vietnam. There are so many intelligent and excellent researchers of oceanography there but the budget is very limited and they do not have ocean-going vessels. Therefore, their scope of research is very limited, only down to 20 m from the sea surface, and they are faced with many difficulties. In the South China Sea, there is no permanent observation station, so important data is lacking from that particular area. Strong research institutions of the world have to make sure to cover these areas where study is still rather behind and lacking. Of course, we have to respect the intention and policy of the government, but the ocean is continuous. There is no physical border dividing us from one part of the ocean to another part. There may be some difficulties in establishing collaboration, however. We are seeing common natural phenomena, which are the subject of our study, and I hope that we will be able to do such work.

There are people representing other research institutions here, and if these researchers would like to make comments on deep-sea research I think there are microphones. Please do make comments through intervention. Please identify yourself before you make your comments. Unfortunately, the time is rather limited.

ISHIKURA: My name is Ishikura. When JAMSTEC was established, I was the president of JAMSTEC for 7 years and I am now the president of the Japan Deep-Sea Technology Institute.

I was listening to what the panelists had to say. Facilities and instrumentation are extremely important for deep-sea studies in the coming years. The deep-sea drilling ships and submersibles which go down to 10,000 m, the manned submersibles, these are quite important. We have to also look at the technological background of deep-sea development. When I was with JAMSTEC, I became aware of government research technology. We also have to take advantage of the engineering technology of the private companies, private industry. The Japan Deep-Sea Technology Association has associations with strong shipbuilding companies, electric power companies, construction, maritime companies; we have participants from these private companies. We are a group of engineers and technologists, and if we can be of any service to you we would be more than happy.

When we look at engineering, we know a lot about hardware. However, the user's requirements are rather difficult for us to identify, and that is one of the greatest difficulties for us. Therefore, the Oceanographic Research Institute is kindly invited to give their users' requirements to our institute so that we can be of more service to you. Thank you very much.

NASU: If I may, I would like to ask Dr. Kitazawa from the IOC secretariat to make a comment. Dr. Kitazawa, would you mind?

KITAZAWA: I would like to have some more time before I make any comments.

NASU: How about Dr. Chen from China? Would you mind making a few comments? Dr. Chen, director, First Institute of Oceanography. No?

Then I would like to postpone the floor session to later during the day. First I asked each panelist to restrict their comments to 10 min, and some of the panelists may like to make supplementary remarks. I can give you 4–5 min if you have any additional remarks. Is there anything that you would like to add?

DORMAN: If I may, I would like to just perhaps continue our previous discussion on the possibility of the workshop so that perhaps more discussion amongst us may lead to some specific issues for Professor Nakato to take action on. I believe that we have currently appropriate and effective international mechanisms within the context of the ocean drilling program now that our friends from Russia have joined us, thankfully, at long last, to be able to discuss the future potential for the new drill ships, and in terms of both technology and the mechanisms by which we may internationally approach that.

I would suggest then that the workshop that you have proposed here, Dr. Nasu, may leave that subject safely to that other group. Not that it is not important, but I think that we will continue our already ongoing discussions on that topic, perhaps even in January in Bonn, as we look at renewing our current agreement, which in itself includes very careful consideration and great appreciation of the efforts being made by Russia, Japan, and France to develop new drill ships for possible use when we have extended our operations beyond the utility of the current SEDCO platform. So we are anxious, I think, very anxious to hear about those technological developments, and the venue for that has reasonably been set.

I would like to support Professor Laubier's suggestion that we may profitably look at the technologies, characteristics, and feasibility studies and designs for the ultimate depth, the deepest presence. This is an argument that has been under discussion in the United States for quite some time, and although there is much interest in this type of a vehicle the interest and the discussions to date have not been extremely well focused on the technologies and feasibilities. Rather, we have been attempting only to see if there is adequate interest and concern to undertake such an effort. I cannot, of course, speak for the country, being a private independent institution, but my sense is that our economic situation and the other priorities for our science would never allow us to undertake such an effort unilaterally. I think this is good. I think the realities of the need for collaboration in these types of efforts will enhance overall our long-range capabilities. But I do believe that because of the discussions we have had, we could make considerable contributions both in terms of our scientific interests in these areas, which of course must drive the technology, as well as the types of manned vehicles we have considered and what some of the technological issues are. So I think we could certainly help on that topic.

The other point that I have heard considerable discussion of today that I believe we could probably include profitably within the bounds of a few days workshop are, if I may, the benthic platforms design to be left behind, to stay and remain on the ocean bottom for longer term observations, of which several of us had, I believe, ideas and suggestions. I think it would be profitable if the developers of those types of instruments were able to exchange their ideas, again in the context both the scientific interest that is driving their considerations as well as the technologies that they are looking at.

Certainly, we have had effective collaboration with JAMSTEC in our efforts on ABE, the two of us have talked about that. I believe comparable international collaboration would benefit us all for the other types of vehicles that are designed. Perhaps this may even fit well with your plans for NEREIS in that I am aware that these are the types of instruments you are talking about deploying from her as well.

So I think perhaps those are two topics that it would be profitable for us to discuss.

LAUBIER: Well, I am pleased to add something after what you have said. The first would be that, in fact, it is very important what you said, that the European project presented by the French people, called NEREIS, is not only a drilling ship. It is also a ship that is supposed to lay on the bottom of the sea. Benthic laboratories of about 5–8 tons heavy things, can be left for months, maybe a little longer, before taking them to the surface again. So it is something more than a light drilling ship. As far as drilling is concerned, the main interest is in paleo-oceanography more than geosciences by themselves. It makes a small difference.

My second comment would be that we have been discussing this in detail, but we can record the possibility to reuse telecommunication cables due to the important progress in optical-fiber cables. A lot of good condition tele-communication cables have vanished in the deep ocean. Some of them are no more than 10 years old, and you know that the average lifetime of a tele-communications cable is about 25 years. You can probably use them with the first observatories to test and observe, to have direct communication with them and possibly to send a lot of waH power needed for the observatory.

My last comment would be that in parallel with this idea of a feasibility study of the present technologies for deep-sea submarines, ultimate depths as I said, I think one of the reasons why one country, even a large country like the United States, says probably we cannot make the decision by ourselves, is also due to the fact that the scientific goals presented by a given scientific community in a country, in only one country, will probably not be sufficient to make the decision, even if the technological answer is appropriate. So I think in parallel to this feasibility study, it would be interesting to have a workshop of leading scientists interested in what is to be done in the trench because this is a question we now have in mind. I know that it is a very small surface of the world ocean, of course, less than 1%, but there are so many important critical mechanisms functioning for geophysical and geological science. There are also some very important questions on biochemistry and enzymology with regards what is going on below 6,000 m. It is enormous but there is a kind of limit there, that some different mechanism may intervene at deeper than 6–8 km depth. So I would recommend also the idea that in parallel some kind of widely open, let's say symposium rather than workshop in that way, should be organized to see what are the scientific goals of the deepest part of our planet.

NASU: Thank you very much. Might I deviate a little bit, Dr. Yastrebov? I am interested very much in hearing your point, Dr. Yastrebov, about the upper boundary. If an internal wave comes into the upper boundary, I think there will be a disturbance at the slope and the density current will get confused. Have you any comments?

YASTREBOV: In the near-bottom layer, at a height from the bottom of about 100 m, we have a ceiling where there is no mixing of the water in the near-bottom layer. We have discovered that this phenomenon is influenced by the sediment unit on the bottom. When water is going with the current on the bottom, this place on the surface bottom makes turbulence which disturbs the sediment unit. While each sediment unit rises, the energy turbulence decreases, to the height of about 100 m. Turbulence energy in this sediment unit equals zero, and in this place we have eliminated the balance and the water and this layer are not mixed. This is only our proposal, but we must conduct experiments in the ocean with a special station on the bottom and with submersibles.

NASU: I have just referred to this issue, because in case a permanent station is to be set up in the deep-sea bed, then the point Dr. Yastrebov pointed out is very important. I think it is closely related to the tilting or inclination of the sea bottom, the slope of the sea bottom. So I hope that you will make further progress in your work on this point.

Now coming back to the floor, are there any further points from the floor?

KITAZAWA: I would like to thank Dr. Nasu for asking me to take the floor. I am Kitazawa from the IOC secretariat.

Last week, I was attending a conference and at the end of the conference I was informed of this international symposium taking place this week. Because of that I decided to take part in this symposium, and I have found it very useful. I thank you very much for that.

I noticed the following points. Professor Nasu has just made the suggestion that at regular intervals, as much as possible, there should be a forum for exchange of leading knowledge or expertise. I think that that is going to be very important. I believe in the significance in such an effort. Perhaps an annual meeting of that kind may be difficult, but every 2 years or every 3 years would be a very sensible and very useful interaction for sharing the most advanced state-of-the-art knowledge.

Such a regular exchange will focus on the efforts to exchange advanced knowledge. At the IOC there are currently 128 member countries, and I have been thinking about the possibility of those member countries sending their experts immediately and whether their experts would be able to understand any exchange or discussion at such a meeting. I think this is very difficult, because the majority of the IOC members are from developing countries, and the level of expertise of those countries' scientists, though they are eager to expand their research activities, is restricted by a shortage of money. They have budget problems and they want to do something more but they cannot because of the budget problem.

I would like to draw your attention to the fact that there are those countries which are eager to do more but cannot because of budget limitations. They are members of our organization, the IOU. In cases where members from developing countries would like to participate in ODP, how can we support them? How can you make it possible? I would like someone on the stage or Dr. Nasu or someone related to the sponsoring of this symposium to answer the question.

Ocean dives and ODP are two areas where it would be difficult for developing country members to start meaningful global interactions because the technology is too advanced. Therefore, as suggested by some participants and speakers this morning, I think that the worldwide ocean observation system is a more desirable and more meaningful area to be pushed forward through international cooperation and interactions, because that is an area where people from the developing countries can participate comfortably.

NASU: Dr. Baker, is there any comment?

BAKER: Thank you, Mr. Chairman.

I would like to make two separate points. The first one relates to what you and Dr. Dorman and others have said. It seems to me that at this time with respect to submersible facilities, there are a limited number of countries and perhaps even before the workshop those groups could set themselves as an international submersible facility group. They could look at the plans that are existing for the use of the facilities. They could look at the opportunities for sharing, for example, on space. They could look at the needs for training and safety, because certainly that will become an international issue. They could consider applications from scientists from other countries because again it would take advantage of knowing the total resource rather than just one resource. They could perhaps offer very valuable advice to countries that were considering developing their own submersibles. They could also consider the necessary support facilities for the submersible operations.

Now this is something that should be organized perhaps by Dr. Nakato, not by the governments, because I take Dr. Dorman's point that if you involve governments there will be so much red tape that the submersibles will never come to the surface.

On the second point, which is completely different, we often are concerned with databases and we say it is too difficult because there are so many data already available. But when we look at what Professor Karube and others are interested in, in looking at the new forms of life on the deep ocean, we have almost no information. I think the group that has been assembled by JAMSTEC should come up with a strong recommendation that wherever organisms are collected in the deep ocean bed or the deep ocean waters that we immediately set up an internationally accessible database.

Whereas Professor Karube is concerned that Australia and France and other countries have their 200 nautical mile zone, I am sure that all of those countries would be delighted to have their organisms listed in a database. So I would hope that sometime in this meeting we can make a detailed submission on establishing the need for that database. I would appreciate the advice of the experts, but I would suggest that all organisms below, say, 1,000 m would be very easily described today. Thank you.

NASU: Thank you very much. The time is coming to a close. I would like to express my sincere gratitude to the panelists for a very successful meeting, and also I would like to express my gratitude to the organizer, JAMSEC. This has been a wonderful chance for us to study this kind of deep study, as it were. Lastly, I would like to thank the interpreter. It is very hard work, I know. Thank you very much.

This concludes the first day of the symposium. Thank you very much for staying with us for such a long time. We will start the second-day session at 9:30. Please leave the headset on your seat.

From 5:30 on the 12th floor, there will be a reception and we would like to have the pleasure of seeing you at 5:30 on the 12th floor.

Earth's Environments and Oceanic Observation

Panelists: DAVID I. ROSS, MARVIN K. MOSS, ZESHI CHEN,
JOSEPH T. BAKER, YOSHIYUKI NOZAKI, SHIN-ICHI ISHII
Session Chairman: SUSUMU HONJO
Rapporteur: TOSHIYUKI NAKANISHI

KAORU MAMIYA: Session 5 is the last session of this symposium. The theme for this session is "Earth's Environment and Oceanic Observation," and will be chaired by Dr. Susumu Honjo, a senior scientist at Woods Hole Oceanographic Institution, United States. Dr. Honjo specializes in geology and biology, his research has an international scope. Dr. Honjo assisted JAMSTEC in an Arctic study.

SUSUMU HONJO: Mr. Mamiya, a very kind introduction. This is a session on "Earth's Environment and Ocean Observation," which is obviously a very large subject area. We are almost talking about the entire discipline of oceanography, so that we might just focus on global change and the role oceanography can play in that change. That might be a little easier to handle, as well as it being more appropriate as you have heard from the Professor Kondo who gave a very elucidating talk just a few moments ago. We will not necessarily follow his steps, but probably in what he was talking about, a very important message was that the most important thing we have to worry about is the global change in relation to the increase of carbon dioxide levels in the atmosphere.

This panel, this distinguished panel, is as you see it here. We have Dr. David Ross, about whom I would like to give a very short introduction as you already know him quite well. He is the director of the Atlantic Geoscience Center in the Bedford Institute of Oceanography (BIO) of Canada.

Next to him, on the left-hand side, is Dr. Marvin K. Moss, who does not need introducing I believe. He has just given a very illuminating talk about his institution. He is the associate or deputy director of the Scripps Institution of Oceanography, which is a very big and august institution on the west coast of the United States.

Also I do not think I need to introduce Professor Chen. He is the director of the First Institute of Oceanography of China, who gave a very elaborate talk about his activities just about an hour ago. He is going to be our panelist.

Behind him we have Mrs. Kyo, who has been our interpreter, and she has very kindly volunteered to explain a few things if Professor Chen has slight difficulty with communication. So you can talk from the floor — this would be appreciated.

It will not make any difference for us if you talk in English, Japanese, or Chinese, which is very nice for us on this panel.

We have Dr. Joseph T. Baker, from whom you have heard about his tremendous activities in the Australian Institute of Marine Science, where he is conducting an excellent directorship. He will also be a panelist.

Then next to him is our friend Dr. Yoshiyuki Nozaki who we call Yoshi. He is associate professor of the Ocean Research Institute of the University of Tokyo, Japan, and his expertise will be a very important constituant of this panel.

Next to him at the end of the table is Dr. Shin-ichi Ishii, who is executive director of the Japan Marine Science and Technology Center, JAMSTEC, of course. He is executive director, which is probably equivalent to associate director in charge of research in the United States, and he is going to present the side of our hosts. We are happy to have him on the panel.

Finally there is Dr. Nakanishi, who has the very important job of rapporteur of this panel, which will make me honest as to about what I am talking about. For anything which has to be filed for the permanent record of this panel, Dr. Nakanishi will record it for you.

Well, I think we have a very short time and a long time, I do not know which to say. Our panel has to end before 17:50, which sounds like a lot of time, but the subject area is awfully big so we have to hurry.

In explaining a little bit about how we can try to achieve the goal today I would like to ask for the cooperation of the panelists, who all have opinions, in only giving a 10-min talk in the first stage. We will go all the way through from Dr. Ross to Dr. Ishii, and after that there will be 10 min of deliberation, probably as an open discussion, where we may well fight or whatever. We will try to reach some conclusion and have some agreement. I hope we will be very relaxed here; we are not going to have a very ceremonial panel, but we will focus instead on being productive this afternoon.

I would like to speak, for a very short moment, about the sort of thing we have been thinking of with regards cooperation on this panel. This is sort of the main theme or problem, and I think before we start we need a discussion or some introduction of the problems.

It is in fact very simple. The most awesome problem we, mankind, faces now is global warming, and, of course, we really have to carefully and rationally face that problem. In this corner of the universe organisms, like ourselves, have made a perpetual cycle of life for almost three billion years on the mercy of the water and the climate, which is moderated by the water itself. That balance has allowed us to survive.

Hydrocarbons have accumulated as the precious surplus of the biogeochemical cycles ever since the start of life in the ocean three billion years ago. Within just 150 years we, human civilization, somehow decided to use significant amounts of the hydrocarbons that life has accumulated as energy to sustain our lifestyle.

One hundred and fifty years is a fraction of a fraction of a fraction of the long history of the earth and the origins of organisms. This activity is in combination with other major activities, such as deforestation, the cutting of forest lands especially in the last few decades. It seems that the burning of fossil fuels, deforestation, and all that sort of thing, has accelerated and there have been increases in the so-called greenhouse gases. This has a tremendous effect in a relative sense on this earth. We are emitting large amounts of carbon dioxide into the air, as you have just heard, just about an hour ago, from Professor Kondo who showed us a chart on a slide. This is very clear evidence that the greenhouse gas carbon dioxide is increasing every year.

Obviously, we have very good statistics on the emission of carbon dioxide gas just by calculating from industrial statistics. Simple kinetics and thermodynamics lead us to believe that these increases would cause elevation of world temperatures. Is this really happening? We have been measuring for too short a time to be really sure. Some modeling shows there is a temperature rise, some modeling does not. Some winters have been quite warm, but Scandinavian people report that they have more glaciers on the mountains of Scandinavia than before. On the other hand, in Sanbanmoora [??] this year there have been much more southwest monsoons than previously. It is therefore very difficult to assess whether we really have a greenhouse effect. The modelists are very busy trying to decide.

Recently, every academic study, such as about the greenhole in Greenland, has tried to find out about the bubbles which have been accumulating for 40 million years. A very small bubble has been discovered which has been encased and beautifully protected for 40 million years. With current technology, the assessment of its carbon dioxide contents has shown that there is a stunning correlation between the carbon dioxide level in the air and the temperature. We are now starting to believe that the elevation of carbon dioxide in the air most probably does increases the temperature of the atmosphere on this planet.

However, the time scale is very important. We are never very sure what is happening yesterday or will happen tomorrow, even with our tangible, five-year southern oscillation cycle, the El Niño-type of thing. Probably the increased carbon dioxide *is* operating to cause warming despite all the terrestrial, astronomical evidence that shows we are now getting into a glacial age.

The missing piece of information is the ocean. Air is very easy to handle because it is fluid and gaseous. The ocean, being liquid, carries a tremendous amount of latent heat. The temperature exchange between the air and the ocean is not simply a one-to-one relationship. It is a very complex [C-I-C] exchange. Heat moves around, conducted not only by water but also by salt and other things. Heat is stored in the ocean not only on the surface but also down to the bottom. The time scale is very complex. Without an understanding of the thermal dynamics of ocean circulation and air/water exchange, we will never know the full picture. The ocean takes a long time — is very sluggish. We may not even have any greenhouse effect. This is probably not true, but it is possible.

Another thing which we do not understand is the behavior of carbon dioxide. We know that we have about six billion tons of carbon dioxide emitted, but only half has shown up, as illustrated in the chart which you saw a few moments ago. Where has it gone? Also, we don't understand how the carbon is operating.

Essentially, we know that it comes from chemical erosion of the land, and then goes through a very vigorous biological stage of chemical modification and then ends up as bicarbonate at the bottom of the ocean. That bicarbonate probably ends up taking part in volcanic activity, such as near Japan, the subduction zone, and comes up the volcano, back again into the atmosphere. That is probably the longest cycle. But in the short cycle, which is a few seconds long, carbon dioxide is utilized by the quick-and-dirty guys, like the phytoplankton; those guys really make a very fast job of the fixation of carbon; then immediately they oxidize and go back into the carbon dioxide. It is a very complex system. We have to understand this cycle of material which we call the biogeochemical cycle.

These two systems are operating and they are very complex. Obviously we have to know everything about the ocean, and from that context, I would like to ask the panelists to give their opinion. First, Dr. David Ross, from the BIO.

DAVID I. ROSS: Thank you, Mr. Chairman.

My talk is about ocean observations. The Purpose of ocean observation R&D is to understand global environments, I am going to concentrate on the issue of global climate change.

First, a few facts. Global climate change has occurred in the past. We have very good records of significant changes in climate throughout the history of the earth and, in particular, in the last few millions of years. It will certainly happen in the future again. There is not much doubt about that. I think we have heard this afternoon as well how there are various ways in which humankind has influenced the climate by changing the concentration of greenhouse gases.

Other thing I want to start off with is that the climate regime has two major components: a fast component that is controlled largely by the atmospheric heat engine and a slow component that is regulated by the global oceans. It is particularly this understanding of the deep-ocean systems that we really need to look in terms of the issues of the longer term global climate change.

The World Climate Research Program (WCRP) has as its first objective to try to understand the extent to which climate can be predicted, and secondly to try to understand human's influence on climate. Such a program clearly requires a quantitative understanding of the physical climate system and, in particular, the four major components of the system: the global atmosphere, the world oceans, the cryosphere, the ice caps and glaciers, and the land surface. In our discussions over these last couple of days we have been concentrating very much on the world ocean systems. But while we have been concentrating on the world oceans, I would emphasize that any climate prediction must take into account the factors from each of these four major elements of the climate system.

Now the WCRP includes all internationally coordinated research activities that study the physical climate system, but it particularly includes the programs that you have heard about already: the Tropical Ocean and Global Atmosphere program (TOGA), a program to study the southern oscillations in the tropical waters and the associated El Niño; the Global Energy and Water Cycle Experiment (GEWEX), looking at transport and exchange within the atmosphere itself; and then the World Ocean Circulation Experiment (WOCE), which attempts to look at simultaneous observations around the globe, but in particular, addresses ocean circulation on a basic cycle.

The first two of these, TOGA and GEWEX, to a large extent address the issues related to the rather short period part of climate variation, those variations of the annual fruit of the decadal periods. In contrast, WOCE has tried to initiate studies that address the longer term factors: the decadal fruit of the century and several centuries, period. As I indicated earlier, I think the issues that we really want to look at much more in the future is the how to obtain information for these broader global study.

The Second World Climate Conference met in November, 1990, and it came out with a clear statement that there was an urgent need to create a global climate observing system. It looked to this global climate observing system within the framework of an improved weather watch program, a global ocean observing system, and a monitoring program with key *climate* system components. Certainly very much incorporated into that is the whole issue to handling the enormous amounts of data that will be part of this and developing sophisticated modeling that can use this data.

So a global ocean observing system is defined then as a comprehensive, integrated system for collecting, analyzing, and distributing physical, chemical, and biological data from our oceans, from our coastal zones, and the enclosed seas. I would like to emphasize that while the key opportunity here is driven by a need for improved climate prediction, certainly a global ocean observation system will provide us with an opportunity for collecting information of much broader interest than just climate prediction. In particular, it would address issues like understanding the biological processes, such as primary production in the oceans, for example.

In looking at what needs to be part of the global ocean observing system, one should consider what we have at the present time, what is part of our present observation system. I just have a couple of overviews here that I would like to quickly go through.

These provide in some way a history of ocean observation, starting back in the latter part of the nineteenth century with the voluntary observing ship scheme, which all nations are very much a part of. Presently we have some 7,000 observing ships providing data on a daily basis back into our environmental and climate programs. Then around the mid 1950s and 1960s, the Integrated Global Ocean Service System (IGOSS) was established, and this was built around the specific capabilities of some 200 research and merchant ships to provide subsurface measurements as well as the surface measurements of ocean conditions. In the mid 1970s we developed a drifting buoy technology which has now become a major part of the programs, and Dr. Moss earlier today talked about some of the new buoy systems. We now have approximately 600 drifting buoys, provided by some 11 countries.

There is also the global sea-level observing system, an important component of climate forecast for looking at sea-level change. There are at the present time some 220 operational stations of a total component of 300 permanent stations that are planned to be in place within the next half decade.

Exchange System plays an important role with regards information coming into national and world data centers for use by all involved in global prediction. The International Oceanographic Data and Information was established some years ago and is a critical part of any observational program.

Since the late 1950s we have had an increasing number of satellites that provide us with a lot of new information on the sea surface and also provide us with many opportunities for the transfer and communication of data from moored systems. A key point here is the suggestion that by about the year 2001 we will be looking at a greatly increased volume of environmental data coming from our satellite data; the prediction here is of two orders of magnitude. So certainly one of our significant challenges is how to use this data and how to provide it for modeling work.

One of the things that is perhaps important from our present observing systems is that for the most part we are looking at interactions on the surface of the sea which impact the relatively shorter term predictions. The challenge, I think, with the development of new global ocean observation systems is to develop and put into place the equipment and technology for measuring the full ocean section. The kind of innovative technology that Dr. Moss showed you this morning on some of the work from Scripps is a critical part of any activity of this kind.

I think it is also very important that, in considering how we develop the global ocean observation system, we look very carefully at the needs of the program, the data requirements, and that we do not fall into a potential trap of developing new tools for the sake of developing tools. Any global observation system is obviously going to be a very expensive tool to make, and so we must make sure that it is built on good scientific plans, that it is dynamic in the sense that it can be changed and modified as our knowledge of the ocean changes, and that we can therefore make this a very practical application system. I think the other important point to make about this is that clearly the development of such a system requires very full international cooperation. Through the various international committees and science plans, we need to be able to develop the international consensus on what is required and the types of equipment that are best suited for putting everything in place.

I think we are having difficulty at the present time in terms of trying to define the requirements of the global ocean system with any real accuracy. Perhaps one of the opportunities we have here is to look at paleoclimate work to enable us to look at past climate events and to try to use this to give us some handle on the time scales and spacial scales involved in some of these things. In finishing off I would like to quickly refer to a couple of slides. As a geoscientist I think one of the opportunities that we have in developing global ocean observation systems is to use the paleoclimate record from which we can obtain considerable detail to use marine sediments to try and address the issue of time and spacial scales. This is just a record of sediment layering that we see in marine sediments, which can provide with us a very detailed and very complete record of climate change. This is a slide of the long coring facility that is being developed jointly by the University of Rhode Island and Canada. It has been implemented as part of the recent Arctic cruise across the Arctic Ocean to acquire cores that could be used in this way.

Our sediment record provides us with an opportunity to look at both high-frequency and low-frequency climate signatures for analysis to look at things like sea-level change. It also allows us to·look at the chemical interactions on the carbon flux. This is an important means of controlling the requirements for ocean modeling and also a very useful tool for perhaps checking our models again.

Finally, just as an example, I would like to describe some recent work that has been done on the deglaciation of North America in the period 12,000–8,000 years ago. I have two slides here from this selection: 11,000 years ago, you can see the beginning of the major runoff from the Hudson Bay and the eastern North American area during the Younger Dryas period. There is a very rapid transition, and deglaciation that occurred during this time with a major impact on the ocean circulation in the North Atlantic and the Labrador Sea. Two of the things we have been able to show from this are that a simple model of a major event that occurred at this time was not quite so simple as one would like to look at, and the very significant impact of changes in ocean circulation both resulting from and causing the deglaciation during this period.

Thank you very much.

HONJO: Well, thank you very much, David. That was most appropriate. We are very lucky to start this panel with such an elucidating talk. Dr. Ross has been emphasizing that basic research such as paleo-oceanography is indispensable, and also he touched on GOOS and so on which was very nice.

In addition Dr. Ross said in his deliberation, that the satellite is very important. The new technology is not now a technology, but a science. The science itself has big ramifications, and my understanding Dr. Moss will be talking about these ramifications from the view: point of oceanography and in relation to the changing of the climate.

Dr. Moss, please.

MARVIN K. MOSS: Thank you, Mr. Chairman. I was asked to talk about satellites and some of the global views on global oceanography.

The first point I would like to make is on some of the words we have heard in the last couple of days: global environment, global change, global climate change, global ocean observation system. They are all global, global, global, and so we are all concerned. One of the best ways to take a look at anything global is with satellites. Satellites do this every day, over and over again. That is the first point that I want to make.

The second point is that in his lecture yesterday, Dr. Papon showed scattering data from a scatterometer on ERS-1, which is very necessary to deal with ocean currents. To deal with ocean currents, you need global circulation models to tell what climate change is really like. You need to know the circulation so it is very, very important.

He also showed some SAR images from satellites, and I was trying last night to recall whether it was Dr. Kobayashi or Dr. Karube who also showed a satellite view of an area of northeast Japan, up close to the location of the Otsuchi Marine Research Center of the Ocean Research Institute, University of Tokyo, showing temperature variations, some eddies, gyres and so forth, off the coast of Japan. To end my second point, I would go back to a quote that I took down from Dr. Papon, that satellite measurements are now coming to maturity.

The third point that I would like to make is to just briefly show you, if I could, some of the things that can be observed from space just to get a common appreciation here together.

Here we see the AVHRR sea-surface temperature map of southern California. There are a lot of these from southern California because we have a satellite facility at the Scripps Institution of Oceanography and we take these many times every day. The clouds are in white. There are two circular objects marked A and B which are meso-scale eddy surface structures, and you will notice a lot of filaments coming off the California current down along the coast of California.

This is again an AVHRR sea-surface temperature map of a large region of the west coast of the United States. The filament structure is ubiquitous. This is very important to acoustics, the acoustic structure in the ocean or destructure, and to commercial fishing.

The next slide is interesting. If you will look from top to bottom, you can see a time series from the 1st to the 8th of July, 1985, in approximate 4-day increments, showing the development of a meso-scale eddy dipole with strong frontal structure in between. If you look at the bottom you can see a structure here and a structure here that are evolving. It is a dipole eddy, which again tells us a lot about what is going in the ocean.

The next viewgraph is a CZCS (coastal zone color scanner) image of the coast off Baja California. The red and yellow colors there are associated with high phytoplankton pigment concentrations. At the bottom are some smaller panels showing the mean seasonal patterns and phytoplankton pigment concentration computed from 7 years of CZCS imagery. But I think you begin to see that there are a lot of things that can be observed from space, even though most of it is not just within the surface layers.

Continuing on, the next slide shows the date there, January 1982 and January 1983. If you remember, there was an El Niño then and you can see the warming trends off the coast of California. Again that is all I wanted to point out with this.

This slide is rather interesting. If you start with Panel B, which is the second from the top there, you can see a typical year showing upwelling at the equator. Around the equator is the blue-green band of water right here, which is normal. This is from 1981. The panel below is 1982, which was an El Niño year, and you will notice that the cold water upwelling is no longer there. The next panel is just the difference between these two, pixel by pixel, and of course you notice here that the temperature is $3-4°C$ warmer in the El Niño year. Again I am trying to illustrate to you some of the things that we can tell about the ocean from satellite observations.

This is again off the coast of California but both of these images are related to sea-surface temperature and to phytoplankton pigment concentration.

The next viewgraph is again just for interest. If you can read that on the left, the red is fish per boat per day. The red is 101–200, green is 51–100, and blue is 0–50. Compare this with the filament structure over on the right, along the coast of California.

The next viewgraph is of sea-surface wind temperatures. This is one of the NASA viewgraphs, but composed from satellite observations. As I said earlier, we need these in order to derive ocean currents. We need the ocean currents to put into global circulation modelling, that we do a lot of at the Scripps Institution of Oceanography, to predict what is going to happen if we do continue to increase CO_2 into the atmosphere.

The next viewgraph, which I threw in just out of interest too, is of sea-surface topography and ocean bathymetry. Compare one with the other and ask yourself which is the bathymetry, which represents altimeter measurements from sea. Of course, you can easily tell the difference there but nevertheless the features from altimetry from the satellite are very, very similar and correlate with many of the things we see from bathymetry.

This slide represents change in sea-surface temperature in the El Niño area, and this is 1987 minus 1988, 1987 being an El Niño year. You notice the warming in the El Niño area anywhere from 1–4°C in this area.

This is 1987 minus 1988 showing change in the atmospheric greenhouse effect. In other words if there are no clouds there will still be a greenhouse effect due to the CO_2 and other greenhouse gases in the atmosphere. This is the measured change from 1987 to 1988 in the atmosphere greenhouse effect. I say measured because this is ERBE's data, the Earth Radiation Budget Experiment. It is a satellite that is used very much by Dr. Romanathan and others at Scripps and was actually put into space by astronaut Sally Ride, who is now head of the division at Scripps, our California Space Division that Dr. Romanathan works in. Actually one of my former graduate students built the satellite. It is really incestuous there.

When one has clouds, there is an albedo change at the upper portion, well into the atmosphere. If you look at the albedo change, that is the reflection of the clouds, you notice a decrease in the W/m^2 by up to 60 with the clouds. Backing up to the previous slide on the greenhouse effect, when there are clouds as compared to when there are no clouds, the clouds are going to absorb more of the reradiated, Stefan-Boltzmann-type radiation that comes from the earth and the lower atmosphere. There is a greenhouse effect due to the clouds in addition to the normal atmospheric effect. The albedo change, and this one is very positive, reflects sunlight back out into space, as picked up by the ERBE satellite, and so it is mostly a negative effect. Add the two together, pixel by pixel again, and you notice that most of the color is in the negative range. So at least in October 1985 over this portion of the earth, the clouds had a negative effect, or a damping, of the greenhouse effect.

It was mentioned yesterday by one of the lecturers that this is one of the big unknowns in global warming. It is a very hot research topic that we really need to understand. None of the global circulation models have any cloud effects built in today. We hear these numbers, and I am not trying to say that it is a negative effect all the time, but it is just a big unknown.

Let me just summarize here quickly. We certainly know that environmental concerns are very real. The ocean plays a very fundamental role as we can begin to see from our talks here today and yesterday. We have a lot of satellite data, but not a lot of people are doing a lot with it. There are going to be more satellites in the future. ERS-1 is up and working. Sea WiFS will be coming on in the United States before long. There has just been a restructuring of the EOS, the earth orbiting satellite program, in the United States. It is going to go from a US$30 billion program down to a US$10–12 billion. Rather than have big, massive satellites, we are going to have a number of smaller payload satellites that we really need. I think this is a really positive point.

We at Scripps are taking this very seriously. We have had a satellite facility for some 12–15 years. We are going to expand that into the future with the other satellites that are coming on because we think it is because it is going to be another important aspect, just as, for example, deep-sea diving is, to the overall understanding of the ocean and the role of the ocean in global change.

HONJO: Well thank you very much. That was a very beautiful illustration, and I am glad you touched on the clouds, which few understand although this area seems to be very important. We have noticed that for a long time we have had, most of the means and strategies apart from the satellite all in the sum for the science of oceanography. Now the total sum has really changed our concept of oceanography and we can see the whole thing in a the synoptic way. So your talk was very important. Thank you very much. I am sure that we will be asking you to talk more in the second round.

But now I would like to ask Professor Chen to take the podium and talk about his proposals. As you know, China is a very, very important area. As Dr. Ross mentioned about the GOOS problem regarding the marginal sea, the inland sea, and so on, China is one of the most important locations in the world for great rivers, joining rivers, and so forth. We would appreciate your presentation, Dr. Chen.

ZESHI CHEN: Mr. Chairman, I would like to introduce China's in-situ observation work as a sub-program of WOCE, or WHP (WOCE Hydrographic Program) in which the Kuroshio region and the Western Tropical Pacific represent important study areas.

1. Now, please look at Fig. 1. PR-1, PR-7, PR-18, PR-19, PR-20, PR-21, PR-22, PR-23, PR-24 — all these sections represent important locations from the standpoint of studying the China Sea and the Kuroshio. They are also

included in the international program of WOCE. CP-4 is part of Section P-4, and is used for WOCE, being situated west of Long. 165 degrees E. Furthermore, P-18 and CP-13 are between Lat. 20 degrees N and Lat. 20 degrees S, and included within Section P-13, and used for WOCE. P-24, P-25, P-26, P-27, P-28, P-29, and P-30, all of these sections plus P-8, P-29, and P-30: they are located where sea water exchange takes place between the Pacific Ocean and the Indian Ocean. It is required that we conduct all-strata (from sea-surface to sea-bottom) hydrological observation at least once during the survey period. Measurements required include CTD, XBT, drift buoy, ADCP, and meteorological parameters at the sea surface.

2. Flux observation. The WHP observation work includes sea-surface heat volume, movement, water vapor flux, vertical distribution of wind, temperature, and humidity at the air-sea boundary layer, sea-surface roughness, and solar radiation.

3. The WHP observation work also involves surface sea-water chemistry and measurements of trace elements. Temperature, salinity, O_2, turbidity, and nutrients are measured at all stations simultaneously with other items for WHP observation.

4. Sea level observation. The Chinese Committee of WOCE is going to conduct sea level observation at Dalian, Yantai, Xiamen, and so on, and the Great Wall Station in the Antarctica.

5. A satellite remote sensing program is also planned.

Thank you very much.

HONJO: Thank you very much, Professor Chen. We are very happy to hear about your programs that are making a very important contribution to understanding by others of this particular area of the west side of the Kuroshio and the way into the East China Sea and South China Sea. We will probably be asking you later on for some more comments, but at this time I would like to ask Dr. Baker of AIMS about his opinions on the theme we are discussing. You heard this morning a very interesting talk by Dr. Baker about his emphasis on the near shore, processing, and measurement and observation, especially in light of the carbon cycle. As we all know now, the coral reef, for example, makes an enormous impact and to understand what is going on, it might be necessary to do more than what we have been doing in research into the deep sea. Moreover, just as in the photographs and photomicrographs shown by Dr. Ross, coral rows are very important for an understanding of the historical view of carbon dioxide evolution on this planet. So we will be very appreciative if you could give us your thoughts on coastal and coral reef oceanography in light of the carbon dioxide problem. Thank you.

JOSEPH T. BAKER: Thank you, Mr. Chairman. The chairman has given me a fair degree of freedom in talking on the topic of the earth's environment and oceanic observation, and in many ways I look at the problem from the fringe of the deep ocean to the shallower waters and where it impacts on society and the interface with the land.

Predominantly, as I said before, we are looking at reefs and mangroves, and we believe that we have an important part to play in the global ocean observing system, not only by satellite but by other methods of remote sensing and particularly where we have the opportunity for accurate and repeated ground-truthing. Again, this ground treating is predominantly at the fringe of the deep oceans, and we are particularly well-placed to look at the impact of deep oceanic phenomena in the earth's environment.

Now before moving onto some of the topics, I face peculiar problems. The public concern for global carbon dioxide levels and global warming is one that has directed both benefit and criticism to scientists and technologists. On the one hand, we are expected to explain the phenomenon and to find ways to reverse or minimize the effect, and in that way we are praised if we are successful. But on the other hand, we are criticized because we caused the carbon dioxide increase and we did not know how to control our science. Now I face this quite a lot in Australia, and I will refer to it again in a few moments.

Many of the matters of concern in tropical regions, notably in coral reefs, mangroves and seagrass beds, are comparably concerned with other oceanic observations rather than just the carbon dioxide level. For example, sea level rise for coral reefs and for mangroves is a matter of considerable concern. In summary, I would say we believe that the coral reefs will do much better than you and me. They are adapted much better.

We are very interested in changes in sea conditions. For example, if you tell us there is going to be increased storm activity or different types of wave activity we would like to put those into our models of the impact on coral reefs. We have established throughout the Great Barrier Reef a system of weather stations on a grid much finer than anyone else has available. We are equally concerned with sediment transport, with nutrient transport, with toxic transport into the sea.

One thing I would like you all to think about, perhaps more than you have to date, is the amount and the wavelength of the light that will reach our marine organisms. We are concerned with the impact of the destruction of coastal features as a part of the earth's changing environment because we have shown that this affects both the biological productivity and the physical protection in the coastal fringe.

Now many of the things that I have mentioned are interrelated. We cannot study the impact of any one without considering others. Certainly for corals, we can show that the temperature, the light, and the nutrients are very significant.

We are now being challenged to advise government on the health of the reef, notably the Great Barrier Reef. Perhaps in your locality, you are being asked to talk about the health of the inland sea or in other countries confined waterways and their ecosystems. At present, we cannot do this; as a basis for long-term monitoring programs we are in the midst of a series of discussion groups in our institute trying to work out what should be measured, when and how, and how often.

To compare it to when you and I go to the doctor, do you know what is the pulse of the Great Barrier Reef or a coral reef, what is the heart, what is the lung, what is the kidney, and what organizations represent health? We have mentioned how important it is to understand the past. Globally, from what I showed earlier today, coral cores represent a significant contribution to the understanding of past global change. There are many more signals that we do not yet understand.

We have to work out what are the best indicator species. Those that are sensitive to changes in temperature, light, sediment, or nutrients, and all those with the widest geographic range, all those which may indicate change by their avoidance of certain habitats, or by their sudden ability to compete successfully. The example in Kaneohe Bay in Hawaii, where sewage was discharged into a very successful coral reef, is a good example of that last case.

We do have to model water flow in complex areas. I believe there is a challenge for the instrumental masters of the world to be able to develop instruments within corals. We want to know how water flows within systems that your major submersibles cannot reach. So please consider the other side, the microoceanography of the vast ocean.

I would like now to show a few slides. I spoke about the importance of light, and if there is a change in the ozone layer we are going to see a vast increase in the amount of light in this lower damaging UV range.

This is a slide that I would like you to consider very carefully because it is the downward irradiance and related biological damage response at different wavelengths of light. At north meters, that is, at the surface, you can see the wavelength of light that penetrates to certain depths. You can also see on the other side the wavelengths of light that cause plant damage and DNA damage. For a long time we have suspected that the damaging effects of light do not reach deep into the ocean but, in fact, they do. The changing conditions of sediment into turbidity are going to effect the amount of light reaching organisms. For people considering regenerating coral reefs, you have to take account of the light, you have to account of symbiotic associations.

This is the environmentally damaging UV light, and as we go to progressively lower wavelengths of light we are going to find that organisms will have to adapt to give a greater protection in this area. They may do this by living in deeper water, or they may do it by generating new chemicals.

This is a slide of a healthy reef. This is the crown of thorns starfish. You may ask what that has to do with the topic today. We do not know what causes its population explosions, but it is a feature of the earth's environment and we know that it has been there since the reef existed. It is an efficient carnivore. It exudes its stomach and like a giant vacuum cleaner sucks out the living coral. It leaves a feature of coral which now is no longer living but represents a wonderful substrate for new life to settle upon. One of the things in coral reefs is the challenge of finding a new site on which to live.

Another aspect of changing environmental conditions in the earth's environment is waste from humans, which we have neglected to some extent. This is an example of a suburban, or urban, waterway in which we have caused dramatic environmental effects. The next slide is one borrowed from the United States showing the effect of building too close to the coast and highlighting the comment I made this morning that if we are sensible in our coastal management we can prevent a lot of the expensive environmental damage of man-made structures.

My final slide is one which I would like to think that you will consider even with regard to your deep oceans. The so-called natural ecosystems which appear stable are based on our perception of certain characteristics. Whether it is a reef or a rain forest, it is going to have a balance of carnivores, herbivores, planktovores, and detritavores. When we impose change, whether it is carbon dioxide, human waste, toxic materials, sediments, or the like, we may have an ecosystem change from A1 to A2. If time is sufficient, that is natural change, the system will still appear to be in balance. The significant aspect is if humans have introduced change at such a rate that natural ecosystems have not been able to cope. The rate of change of ecosystems will depend on the nature and frequency of impact and the effects of those.

Repeating what I have said and others have said, we must understand temporal and spatial variability of ecosystems before we can understand the effects of anthropogenic impacts. Systemic monitoring is necessary, and in this fringe between the ocean and the land in which I work predominantly we do need to consider understanding of temporal and spatial variability in order to understand the anthropogenic impact. Thank you.

HONJO: Thank you very much. We are very impressed by the interesting deliberation, and especially I liked very much your analogy with the anthropogenic demolition of the coral reef to the deforestation of the land. Also what we have been really trying to do recently has been seen a great deal in the news, in the newspapers for example in many countries. In many places the coral reef is not seriously considered for preservation, but it is so important. It is not really replaceable by an airport or some industry, or land, or so on and so forth. I think those are the issues; they are very important in order to not only try to understand the CO_2 but also the coral reef, which seems to be an extremely important system to preserve our present level of the CO_2 as much as possible.

Well next our speaker is Dr. Nozaki, and as I said, there are two systems we really have to understand. One is the circulation of the ocean to understand heat flux and the interaction with the heat and the air by the greenhouse effect, and after we understand how the ocean is heated then we may understand what is going on. The other side of the coin is that we really have to understand the recycling, or biogeochemical cycle, of the problem. Otherwise, we will never understand what is going on. So those two things are most important issues. We would like to talk about that problem, and Dr. Nozaki is talking about his experience about the recycling or circulation or the movement of the material, especially the carbon in the ocean system and how it is related to the study of the understanding of the carbon dioxide problem.

YOSHIYUKI NOZAKI: Thank you Mr. Chairman.

I would like to make some comments on the ocean and global change from the viewpoint of my expertise in biogeochemistry. At present, two major oceanographic investigations are going on, the World Ocean Circulation Experiments (WOCE) and the Joint Global Ocean Flux Studies (JGOFS), both of which are very important because the ocean plays a significant role in controlling the world climate, so exchange of heat, water, and carbon dioxide act as this air-sea interface. There is no doubt that without having detailed knowledge of oceanic behavior, it is impossible to precisely predict future trends of global climatic change which may be caused by human use of fossil fuels. This is a point which Dr. Honjo raised.

So now let me briefly review how marine geochemists have contributed to an understanding of the global problems. Our major concern is the composition and distribution of elements in the ocean. The composition of the major elements was established by the end of the nineteenth century when the British Challenge expedition was made. Since then, major efforts have been devoted to the determination of trace elements in seawater. However, for quite a long time it was not successful for many of the heavy metals because their concentrations are extremely low and the seawater is very easily contaminated.

There is a famous story that a German chemist, Fritz Haber, tried to recover gold from seawater in order for his country to be able to pay for the deficit resulting from World War I. His attempt completely failed, however, because the concentration of gold in seawater was at least three orders of magnitude lower than the volume given at the time of the program of study. There are a number of other stories which show that various reports on heavy metal concentration in seawater differed with time. Of course, these were not real trends but artifacts.

This tendency continued, even when the Geochemical Ocean Section Study (GEOSECS) was initiated in 1970. GEOSECS was aimed at obtaining a section map of geochemical parameters in the world oceans, but it could not include trace elements except for barium because of the lack of analytical consistency. Therefore the problem focused on the radionuclides and stable isotopes.

In 1976 reliable data for some heavy metals, such as copper, nickel and cadmium, because available owing to the careful improvement of techniques. In the following 15 years to date, there has been a quantum leap in our knowledge of trace elements in seawater. I will show you the most recent data sets for the trace element distribution in the North Pacific, according to the periodic table. If I had attempted to make this kind of figure 5 or 10 years ago, I would not have been able to fill one-third to one-half. Even now, there is still a lack of data for some prominent metals, lawrencium and osmium, niobium and tantalum, and so on, but there is a good chance that all the data will be filled in by the year 2000. This is a very basic study in marine geochemistry, but I wish you to understand that it is not only for the satisfaction of human curiosity but also for the solution of various very important oceanographic problems.

In terms of global change, I would like to emphasize the following three aspects. The first is the usefulness in research on ocean circulation of the various chemical tracers of steady and transient states, such as radiocarbon, tritium, helium, and freon. These tracers are useful to estimate the rate of basin circulation, basin-wide circulation and ventilation, and thermocline mixing, which is somewhat difficult to obtain by other means. Some of the trace metals, such as manganese, aluminum and methane, are also useful to survey the hydrothermal events.

The ocean circulation is undoubtedly an important control of global climate. For example, extensive supply of nutrient-rich deep seawater to the surface, increases ocean productivity and thereby reduces atmospheric carbon dioxide. Ocean circulation also carries cold water up to the surface so that the surface temperature may be reduced. Thus any change of ocean circulation can influence the global climate. Chemical tracers provide one useful tool to study such physical oceanographic behavior.

The second point which I want to make is the importance of biological processes. Here I quote John Martin's work on iron. It has been well-known that micronutrients, such as nitrogen, phosphorous, potassium and iron, are essential for plant growth. For terrestrial plants, it is hard to believe that iron can act as a limiting nutrient because it is very abundant in soil. However, in the ocean, iron is highly reactive and has a very short residence time, and therefore it is likely that a lack of iron can limit phytoplankton growth. Martin collected seawater using a clean technique and cultured plankton with and without iron. He observed enhanced plankton growth with iron, suggesting that iron is indeed acting as a limiting nutrient.

He also made careful seawater measurements and found that the surface waters in high latitudinal regions differed in iron, and also some nitrogen and phosphorous were still present. He further argued that during the last glacial period, high latitude ocean productivity may have been enhanced due to an increased wind-blown iron supply. This mechanism may explain one aspect of the atmospheric problem of carbon dioxide concentration whereby carbon dioxide levels were approximately 90 ppm lower than the present-day anthropogenic levels as indicated by an earth core study.

What I want to point out here is that the measurement of iron in seawater was a very difficult one because it is easily contaminated from rust from ship hulls, and this is essentially a real basic study. I imagine that when John Martin studied iron, he did not expect it to take on such an importance in global change. Science is always filled with surprises, and more new ideas, like this, may be generated in the future. But this example clearly tells us that broad ocean science is essential to our understanding of global systems.

The third point which I want to make is the usefulness of trace metals as an oceanographic indicator. In the course of analyzing some heavy metals in seawater, it was found that cadmium is well coordinated with phosphorous throughout the ocean, and from this, Edbert of MIT came up with the idea that cadmium in Foraminifera of deep-sea sediments may be used to estimate various phosphate concentrations in the paleo-ocean.

After development of a cleaning technique, he succeeded in obtaining cadmium and calcium records, which are important constraints on the global geochemical cycling, together with oxygen and carbon and isotope records. One of the important outcomes is the new idea that during the last glacial period the rate of deep water formation in the North Atlantic was significantly smaller than today. Again, this change of ocean circulation may be related to the climatic change, and there are models proposed in this regard.

We are now able to measure chemical constituents of seawater accurately enough to use them as oceanography traces. Future development of techniques will provide more automated, precise, and accurate methods. Using these techniques, we will be able to explore more sophisticated phenomena in the ocean. It has been dream of marine geochemists to obtain a detailed three-dimensional map of geochemical properties in the world oceans. A part of it maybe achieved by the WOCE hydrographic program. However, it is not planned to cover any aspect of the geochemical cycling on a global scale. Probably this could also be part of JGOFS, but due to lack of funding I believe it is now oriented to process study, and so the global description will be left behind. We need to have another worldwide program in the near future which certainly requires international cooperation. I think I will finish here. Thank you.

HONJO: Thank you very much. When you actually discussed the very important aspects of tracers, in terms of ocean circulation and understanding the biogeochemical cycle. That was a very important message for us.

So to the last speaker of the panel. I would like to ask Dr. Ishii about the very important aspects of CO_2, the problems in the ocean, instrumentation and seagoing facilities, and so on and so forth. Dr. Ishii, please.

SHIN-ICHI ISHII: Thank you, Mr. Chairman. I would like to discuss the methods of observation and development of various equipment for future research work and surveys.

First, I would like to discuss the observation methods. Yesterday, as Dr. Asai mentioned, the emergence and availability every time a new technology is introduced has been responsible for exciting breakthroughs and progress in all areas of science, and the same applies to the area of oceanography. The development of new technologies is indeed very important. In the case of the Pacific, which is a very vast and complex ocean, there exists a number of physical, chemical, and biological phenomena with different or varying spatial and temporal scales. Therefore, appropriate observational methodologies are necessary in order to be able to obtain data which are suited to the purpose of each area of research program.

Usually for observation for the purpose of marine studies, what is required is to be able to simultaneously and on a continuous basis observe in real time a wide ranging area, and at the same time real-time acquisition is desirable. Also, obviously, three-dimensional observation will be required. Therefore, in my presentation I would also like to refer to various issues and problems we must tackle in the area of observational technologies.

First, global observation is very effective and as you are aware satellites and aircraft are used for observation. Dr. Moss pointed out that the satellite observation technique or approach has more or less matured. I would on my part like to raise two other issues here. Most specifically, for global observation, satellite data are indeed very important. It is also important that the continuous observation by data is very crucial. For instance, for the purpose of the observation of the color of the ocean, such continuous observation is very much desired. Also, another important aspect is to obtain sea-truth data, and this is especially difficult for instance, in the Arctic. If such sea-truth can be obtained, then the accuracy of the data will be enhanced so the value of the observation itself will, in turn be enhanced.

In the technology of remote sensing, we are actively promoting technology using lasers. As was also mentioned yesterday, overall images can be rendered by emitting the laser toward the ocean surface and into the ocean. If there is any suspended matter in the water the laser is reflected back. The reflected light is observed and measured to observe the suspended particles. We can now measure down to depths of about 50 m, and by using red emission light we can also measure phytoplanktons. Vertical resolution is currently in the range of 1–2 m.

We are working on the following area at present. Such equipment is carried on oceangoing ships and in many different areas sea-truth data and observational data are being compared and analyzed. As a next step this time from aircraft, lasers can be emitted toward the ocean surface and to measure the suspended particles and the amount of phytoplankton on a global scale or other wide ranging scales.

When it comes to remote sensing from the air there are weaknesses, as this measurement is only limited to the ocean surface and the subsurface layer. For underwater global observation what is necessary will be the next question. As a means to achieve this, oceanic acoustic tomography and development will be required. Sound travels easily in water, and we are capitalizing on that idea using an acoustic source and a receiver. We are trying to observe by this method just the time required to travel from the source to the other side in order to understand the structure of the water temperature and also flow rate. It sounds like a nice idea.

Currently, we are using a 200-Hz sound source using a magnetostrictive material. As we are measuring the time required to travel this technology obviously requires a highly accurate clock, but again if the distance varies there will be error. So using a transponder, we are also monitoring how much variation or drift occurs so that the measurement can be made as accurately as possible.

Another feature is transmitting data from a buoy through a satellite more or less on a real-time basis. The data can be transmitted to the observational and analysis center, and it can be analyzed more or less instantly. That is the system we are aiming to develop and establish, and this can be installed so that global observation can be established. Our target is a 1,000-Km^2 area as an observational area or target.

With global or large-area observation, continuous monitoring or observation will also become necessary. With regard to oceanic phenomena, what is required is continuous and time-series simultaneous observation. For this purpose, for instance, manpower-saving maintenance-free observation stations will need to be developed. That would be a quick way to solve this problem. Of course, such as station will be equipped with a number of sensors so that various observations can take place simultaneously. This is important. We will need to install many of these so this will entail quite a large cost as well as a large number of vessels. What is particularly important is to accomplish long-term, continuous observation which, in turn, means we need to improve the reliability of the equipment.

As I said, manpower-saving features are also very important, so development of highly durable batteries will also be required. As for three-dimensional observation, as I mentioned earlier on, oceanic acoustic tomography is included here, but when it comes to deep-sea observation for that purpose, submersibles such as *Shinkai 6500* will be required, as well as the development of deep-layer buoys that is currently going on at several institutions and laboratories. The synoptic data which will be obtained from such observations should be transmitted to researchers via satellite as soon as possible, or in real time, and a network for such purposes will be required.

So for a synoptic time series, real-time simultaneous long-term observation, it will be necessary to organically combine all these values and technologies so that regular or periodic observations can take place. Also need to maximize the efficiency of each technology. Depending on the changes in the environment thorough and detailed observation will be required. Oceanic phenomena fluctuate vastly by region, time, and depth, so observation which is best suited to each will be required.

Another important aspect is the accuracy of data, depending on the research purpose, highly accurate data may be required, and here I am mainly talking about observation by vessels or ships. In the case of WOCE, in order to calculate the flow from the subsurface to the deep layer, it is necessary to measure the seawater temperature and salinity at very high accuracy. Reliable data will be required, and for this purpose establishment of highly accurate calibration technology will also be required. Such acquired data must be made public, so that as many researchers as possible will be able to use such data.

Another emerging major problem into the future is to try to obtain uniform data, standardization of the observation technology and technological cooperation in order to improve mutual understanding will be crucial. Thank you.

HONJO: Thank you very much, indeed. All measurement issues that you have just commented on synoptic, constant and long-term observation, precision, and reliability. All those are very difficult to achieve at the same time. However, in order to understand the global warming problem it is very important to achieve each requirement. On these we cannot compromise: the highest precision, in the longest, synoptic, simultaneous. This must be the goal in development instrumentation for the study of global change.

Coming back to think about the more philosophical points of the observation, one of the contradictions we have is the time scale. We have been talking about faster recording. We have talked about what is probably a very important point in paleo-oceanography, where we once had a different temperature. In the Cretaceous period, for example, about 100 million years ago, the earth's temperature was probably 10° C higher than now, and about 40 million years ago we had a major drop of the temperature and so on and so forth. It is extremely important to understand a longer time scale, but at the same time we have to see the instantaneous every day change of the sea status. We have to solve that problem, or at least we have to have those things in mind when starting the actual research. If I could I would like to ask Dr. Ross his opinion about the time scale, especially the usefulness of the longer time scale research for global change.

ROSS: Yes, thank you, Mr. Chairman. I think the problem of trying to address the time scale comes down to a problem of information theory to some extent in that we have within our climate signal a very significant annual variation, which tends to bury the longer term signal over the periods of observation that we tend to look at.

Just to briefly show some of this, I have a slide showing some of the temperature trends. We have seen this in a number of forms earlier today. I think the important point here is that if one analyzes these measurements from different parts of the world, there are certainly trends there, but there are also very marked variations. When we start to look at the effect of the oceans on these climate trends, we run into difficulties as the longer term fluctuations within the oceans are very poorly known.

In his presentation on global ocean observation systems that John Wood from the United Kingdom made some while ago, he looked at the question of the required precision of measurement of long-term observations in developing any kind of understanding of climate prediction. His estimation was that we really needed to look at something like two orders of magnitude of improved accuracy over the existing measurements if we are going to try and determine the trends more accurately from the existing signals being measured. Now obviously our measuring systems are improving very significantly, but the challenge of determining the noise from the signal in the longer terms is obviously very important.

I think the other point I would just like to make refers to what Dr. Baker mentioned regarding the input from our coastal areas and the input of pollutants from land, which clearly have had an impact on the coastal areas and have created significant chemical changes in the oceans. Pollutants in the sea are going to have a significant impact and introduce a very real additional noise signal into anything that we are monitoring, and I am sure we do not have any particular understanding of the impact of this on the longer term scale at this stage. I would agree with Dr. Nozaki about the importance of looking at the chemical fluxes and chemical processes in the oceans to try to determine their impact, particularly on the longer term time scales.

HONJO: Thank you very much. I would like to ask Dr. Moss about your idea of the time scale, especially how to handle shorter time scales.

MOSS: I would like to show a couple of things here that I have along with me. I was not anticipating such a question, but I think it is a good question. Let me just mention that we have been running a lot of ocean-atmosphere-coupled general circulation models at Scripps, and we use the El Niño region as the "ground truth-in region." It has been pretty well sampled and we know at least surface temperatures and what is happening there. If these couple of models do not really work with such a large signal, they are not going to work with smaller signals in other places, which may be very, very important. So it is a sort of a ground truth-in exercise, [Gadonkin]-type experiment for us.

We have a model of ocean circulation in the region which we run 9 months ahead of actual measurements in the El Niño area. For example, here are the measurements of sea surface temperature over time, going up now to about where we are in 1991. Here is the forecast model that is running 9 months ahead of it; this is an actual run here. Except for in 1986–1987 where there was really a little double peak, the model does not do very badly. That was the point that I wanted to make to begin with.

Looking at a general ocean circulation model for the same El Niño region using present-day, mean surface temperatures, this is sea surface temperature versus year, here is 1950, 2000, 2060, and so it runs for 100 years. Basically the peaks of the sea surface temperature at least up until the present time are not badly correlated with the El Niño events, which is the point that I wanted to make.

Let us look at a 100-year run, from 1950 to 2060, with a GCM using an increase in mean sea surface temperature as predicted by one of the general circulation model runs that you read a lot about in the paper. Actually, this one was Hanson's model, and so the difference here is that we are using the increase in sea surface temperature with a doubling of CO_2 in the atmosphere over 100 years. Now again we cannot really vouch for the validity of the model, but notice what happens right away. Up to 1970 or so and with increasing sea surface temperature, the magnitude of this maximum temperature rise on the present-day sea surface temperature is on average about 1.5°C. There is one prediction that goes up to about 3°C. This is way out of the ballpark. Now the researcher who is doing this is Dr. Tim Barnett, who is a good physical oceanographer and a good modeler, and Dr. Art Miller, a young researcher. This really scares Tim Barnett because the El Niño is a big climate change in the equatorial Pacific Ocean, and it has a lot of impact. It has impacts all the way into the monsoon region as people have mentioned today. It can even be correlated to some degree of validity with snowfall in Siberia. The Scripps Pier that was built in 1915, the Coup de Grae was lost in the 1982 El Niño event. We have a new pier now that was dedicated a couple of years ago.

As I say, this frightens some of our researchers. However, we do not really no how valid that model is, and we need to understand the problem in more detail. There are not any really good coupled atmosphere ocean general circulation models in existence today. That is one of the real problems. We have good atmospheric models coupled very loosely to very simplistic ocean models, and one of the things that we are trying to do at the Scripps Institution of Oceanography is develop good, coupled ocean atmospheric models to run in these GCM-type calculations. We have a supercomputer center at Scripps, and we also use others around the country, in particular the two at the national labs, Livermore and Los Alamos. We work closely with some of the atmospheric scientists there.

Since we do not have good GCMs to run, there is not a lot of confidence in the 2–5°C temperature rise. It may be a lot more. It may be a lot less.

HONJO: That was a very interesting and useful opinion; thank you very much for preparing within 2 minutes. That is a very short time scale anyway. We really need more GCM and other model-based research; we really do not know how the circulation model is working for sure. There is much basic research needed. I would very much like to dwell on this for another couple of hours, because that is the basic ultimate goal here, modeling and a predictable model. However, probably we have to go ahead and talk on another subject matter, that of space.

Time and space is oceanography. To measure time and space with tracers is the simplest and probably the most endurable oceanography research you can run, but at the same time it is not easy. Space is very important. In southeast Asia we have the Himalayas, for example, which give a tremendous impact such as the monsoon, probably influencing 60% of the human population. So in order to understand the atmospheric change, we have to think about the geography and so on. The GOOS project places emphasis on the marginal seas as well as enclosed ocean and open sea. So that is not only time scale, but also geographic space scale.

I would like to ask Dr. Baker what he thinks about the geographic scale and how to handle it. For example, if somebody wanted to immediately instigate a very efficient GOOS-type of project, what would you suggest.

BAKER: Mr. Chairman, I'm not 100% certain that I will answer your question correctly.

With regard to geographic significance, one of the things that you said earlier drew my attention to the fact that Australia has only had farming for the last 200 years. China has probably had farming for the last 10,000 years, and Japan for a similar time. I think there would be value in the geochemical studies looking at the difference of new countries, like Australia where the human impact is much more recent. Now that is not in any way related to your question, I feel, but it is something that has come out of the earlier discussion.

The other point before I try to answer your question is that from what the corals tell us, 1860 to 1990 is too short a time period. It is not adequate, and I would like to know what the people here feel from their measurements and experience, what are the time scales that we should be looking at to detect significant change.

However, in the question that you have asked, with regard to the geographical significance, for example we have found a much greater change in signals in the corals as we move from the coast to the outer reef than we have as we move north-south. So that suggests there is a very significant land influence, even well before the human settlement.

The other aspect, geographically, are that the signals associated with fluorescence and therefore indicating past river flows are only detected on inner corals. When you move to the outer corals, you get no fluorescence, so that you can detect different factors in the outer reef but you cannot detect any river flows. Now that has had a peculiar benefit in looking at corals from the Red Sea, where at the moment because there is no vegetation, they do not have any fluorescence. But we have looked at corals from the Red Sea which are 250,000 years old, and they have significant fluorescence in the fossil coral. So the corals can tell you changes in vegetation pattern well in excess of human historical record.

Now I feel I have done a terrible job in answering your question but I think I have asked a couple of others.

HONJO: We appreciate your comments. In terms of the geography and the space, I think one of the important issues is the inland sea and coastal area versus the marginal sea, which you have been talking about. The very important contribution you made this afternoon that really impressed us was on the importance of the coral reef, which plays an important role in the carbon cycle. We know that many papers indicate the importance of accumulation, supply and recycling of carbon monoxide on the shelf. The shelf, of course, includes the coral reef, but those shelf areas are very unhomogeneous, heterogeneous, very difficult, dynamic areas. There is also the interaction between oceans, continents, and rivers. We have in Asia very gigantic rivers, like the Huang River, Probably the largest seamen's road possible. Also in terms of the carbon input probably a major influence is the Yangtze River, of course. Then we have the Ganges River. I was just wondering, Professor Chen, if you could give us some insight into your program in terms of the GOOS-type of observation, how you see the study of those marginal seas, such as the East China Sea, Which are heavily influenced by the input of terrestrial water.

CHEN: The impact of rivers on the ocean is when fresh water is discharged into the water. There are nutrients and trace elements that diffuse in the ocean and the Yangtze River and Huang River, they have a big impact on the ocean when there is a lot of precipitation in the rainy season. A lot of papers have been published about the effects of big rivers during the rainy season on the ocean.n

HONJO: I think we are now pressed for time, but I would very much like to open this session to the floor. If you have an opinion or statement or anything, please just speak up and come forward, because we are quite relaxed in the open session here and we enjoy your input.

One thing which we have to consider is that this large joint study of the ocean toward an understanding of global warming is oceanography itself. Obviously you see a tremendous amount of data generated and we have to consider how we are going to handle the data. The localization of data may happen because not every nation or even maybe researcher or every institution has the same policy toward data. We ultimately want to open all data so that everybody can use it instantaneously, but of course the problem of credibility as well as the influence of the funding agencies and the situations and the histories are all very complex. At the same time, we really need to work together with not only the advanced countries who have enormous resources, tools, and funding, but also we have to really talk about working together with countries who are not privileged at this moment to be able to participate in the whole thing. We have to worry about some of the social, political, or even the economic problems in handling this kind of large-scale research.

So far, the program we have been running in advanced countries is not necessary for the advanced countries, but in this kind of global scale study we really have to involve every country to be able to include their expertise and their data. Then we have to talk about the exchange of information and the transfer of technology. We need some authority, probably, to ask foreign governments, especially governments of the less developed countries, to participate. That kind of interaction and the philosophy of how to really deal with this large problem is in itself a big problem, and in its implementation there are lots of difficulties and problems. Professor Nasu has been working for a long time to really encourage the so-called WESTPAC programs, and I think that has been reasonably successful. We have to learn these techniques, but at the same time we really need systems in terms of how to run those programs.

We have now Dr. Pugh from Wormley, UK who represents the United Kingdom on the IOC, the Intergovernment Oceanography Committee, if I am correct. I would like to ask what is your opinion about intergovernmental participation.

PUGH: Thank you, Mr. Chairman. You did warn me about this question, but like one of the previous speakers I am not sure that I have the answer. Nor am I sure that coming from the United Kingdom yet qualifies me to speak on behalf of the developing countries. But I do represent the United Kingdom, as you say, on the executive council of the IOC, and our discussions over the past 2 years have been very much concerned with the development of GOOS, the Global Ocean Observing System, and how developing countries might get involved.

I think the first thing to recognize is that if we are talking about scientific experiments, then we should not necessarily involve intergovernmental agencies. We can carry out scientific experiments bilaterally or with two or three countries without the heavy overheads that are involved in going through the inter-governmental United Nations sort of machinery.

If we are talking, however, about monitoring systems and systems for detecting long-term climate change, then I think that permanence is implied. Permanence requires some sort of government commitment by all governments. Then you have to go into the UN system, and that I am afraid is very much a noodle soup. There are all sorts of organizations which have an interest in the sea, including

UNESCO and the Intergovernmental Oceanographic Commission, but also the World Meteorological Organization, which is interested in climate and in flooding and storms. You will have the FAO, which is concerned with fisheries, and you also have the progressively more involved UNEP, the United Nations Environment Program. I think many of us know that next June the UNEP will be meeting in Brazil and discussing the UNCED development and economic development and the environment. The developing countries are very interested indeed in how to strike a balance between those two.

They are interested in the global ocean observing system, but they are more interested in their own coastline and the impact on their coastline. So their interest is particularly, at the moment, centered more on the parallel system which is being developed of a coastal ocean observing system related to climate change. This is something being done jointly by the UNEP, the IOC, and the WMO.

If I might perhaps try to answer quite specifically your question as to how to involve developing countries, I could fall back on a slightly different experience, and that is as chairman of the Global Sea Level Network, which is a component of GOOS and which Dr. Ross mentioned in his introduction. I have been chairman over the past 5 years of development. Developing countries are very interested indeed in coastal sea levels. They are interested because of inundation. They are interested because of erosion. They are pleased that because they measure sea levels it makes a contribution to a global program. But first of all they want to know that it is important to them in some local or regional way.

So I think, first of all, one has to involve them with training and with an exchange of information, involve them in the planning, take their advice. In many cases, they know a great deal of things which we need to know. The other thing is to emphasize the national benefits. We have had no trouble at all in persuading the Maldives and Bangladesh to take an interest in sea level. However, there are not always the government structures able to respond to that sort of requirement immediately so one has to be very patient. I think that is another point.

But always I think we have to emphasize erosion, fisheries, pollution, the sorts of things which have a coastal impact. I think above all else we need to be very positive about their involvement and about the oceans in the role that they are trying to develop. I heard, first of all, this morning that some of our developed estuaries have many layers of pollution, season on season, and I thought Dr. Baker put it very effectively later this afternoon when he said quite simply that these developing countries do not have that sort of problem. It is for us to help them to avoid it. Thank you.

HONJO: Well, thank you very much. From your comments let us start to think very seriously on the level of the difference of the interest in the study of sea level or erosion, coastal area impact versus the open ocean, oceanography. I remember yesterday, Dr. Baker made a comment from the floor about this problem, and I would like to ask you again once more for a short comment from a more practical point of view.

BAKER: I thought that while all the directors were present and while the JAMSTEC directors were also present it might be an opportunity to have the directors meet after today to discuss what we have learned from this session and what we could do among our institutes to develop a system to recommend what we believe should be monitored, to recommend how we can collaborate in the future, and to therefore set the example of sharing information and making measurements in a consistent, systematic, and disciplined manner so that the databases produced will be useful to all countries. Now that may be idealistic and my colleagues will tell me when we meet, but I believe JAMSTEC has provided a wonderful opportunity for us to have the discussions. Somebody said yesterday if we go away without really cementing relationships, we have failed. So I think we will meet after this, and we would then recommend back to President Uchida in the way we think it is wise to move ahead.

HONJO: Well we all share the feeling that if we just run away from this meeting it would be a wasted opportunity. Also we have the responsibility to keep working on the impact by science. This is an enormous opportunity which we have now. We have gathered some very influential people, like yourself, and from learning institutions from the administrative point of view or the scientific administrative point of view. But at the same time the next level of the talk has to be involved with science and some excitement of science versus scientist, so we really need another level or another opportunity to meet.

I would like to propose another workshop following the spirit of this enthusiastic meeting that we have been seeing and the enormous sensitivity, kindness, and understanding of science from JAMSTEC. I would like to ask Dr. Ishii if you could consider another meeting or workshop to follow up this very interesting meeting.

ISHII: Mr. Chairman, you have come up with a very significant, meaningful proposal. So the idea is we should not just forget about the whole thing at this point. Unless we follow up from this meeting there is no way we who are related in the field of oceanography will be able to face the world, as it were, to fulfill our responsibility. So the proposal has been made to hold a workshop after this; if there is any institution or organization which will be willing to sponsor and organize such a workshop I would like to ask that institution to function as a host.

In any event, JAMSTEC is ready to give its full support. After the first meeting, if JAMSTEC is nominated, we at the center will be ready to start making preparations for the second round of meetings.

HONJO: It is a burden on you, but you already have a history of having this efficient and interesting meeting, so I am glad, we are all glad you are willing to accept that burden again.

We have another 5, 10 min or so, and I would really appreciate some opinions and comments from the floor. Would you mention your name and affiliation.

DAN KEDATUM: Yes sir. Thank you, Mr. Chairman. My name is Dan Kedatum from Cinares Headquarters in charge of the ocean and atmospheric sciences at [Cinares] in Paris.

We heard Dr. Moss this afternoon, and we saw some results presented by Dr. Moss on modeling. It seems that modeling is going to be very important in the near future, and it is already becoming so. On the other hand, we heard Dr. Ross telling us that in the near future, a couple of years from now, the amount of data we are going to get from satellites will be about two orders of magnitude higher than what we get now. Obviously, there is a need for interplay between models and data. So I would like to have some comments on this question of the interface between data, satellite data, in fact, and models in the near future.

HONJO: Dr. Moss, would you start, and also I would like you to be backed up by Dr. Ross. So please, Dr. Moss first.

MOSS: Let me comment briefly. It is a real problem, and as I mentioned a little earlier, I do not think we have fully taken into account the observations that we have from space today, primarily for the reason that you mentioned here that we do not have adequate databases. I know that with the EOS system in the United States, which of course will not begin to fly until the latter parts of this decade, maybe 1996 if we are lucky. NASA is already talking about setting up major data centers, and the data will be archived in a way that it is accessible, maybe even by computer, to individual researchers around the country and hopefully around the world.

At Scripps we have been talking a little bit about this and we are actually thinking of putting in a proposal to become a center providing this service. It is a big service function, and we are a research institution; we would rather someone else did it and let us use it, but there are not a lot of people stepping forward on these things. Nevertheless, it is a real problem and it is a problem that has got to be solved. I cannot tell you exactly what the solution is, but we are very much aware of it.

I would like to make one other comment regarding the talk about the data and so forth that we have thus far. I was talking a while ago about a couple of atmospheric ocean models. We need these models, as we all know, for realistic general circulation models. We have good atmospheric models, but does anybody really know of a good single ocean model, a good basic model? We have some models of the Gulf Stream and a few other places, you can get gyres, and you can spin off eddies that the authors feel area really good. However, in fact it is very, very difficult to get a good ocean model. You do not get a good model without tremendous amounts of data so that you can use the data to construct your model and, of course, to ground-truth it. We are not even close to that today; we are going to have to have the kind of data that you mentioned here to do any credible job at all.

HONJO: I think this is the root of the problem, and I would like to ask you, Dr. Ross again, if you could talk about ground-truthing or sea-truthing as well.

ROSS: Let me just go back to the question of databases briefly, and I think we have heard a couple of points that were raised yesterday. Dr. Dorman talked about his technical oceanography, and Mr. MacPhee about operational oceanography. I think we are all trying to address this problem of how to get the data together, how to manage it effectively so that it is available on a broad base. Clearly the world data centers are going to play an important role in this, along with research institutions, in developing the means for handling this data. I think it is a problem that is going to be attacked in a number of different ways by different institutions, and that is probably not a bad idea because perhaps out of that we will all learn how to handle it better.

With regards ground-truthing, I think there is a major difference between environmentally forecasting and weather forecasting where within a relatively few hours or a few days we know how good that forecast is.

Part of our problem with ocean modeling and climate modeling in the longer term is, of course, the problem of how to ground-truth this and how to look at the variations that address this. If one looks at climate prediction as a problem of chaos theory, then I am not at all sure that you can ground-truth it for the future. But clearly, part of the observational systems are going to have to provide us with this mechanism of looking at a control of our models.

I think that the past record provides us with an opportunity to test the importance of the variability within the ocean system and its response to past climate changes. It gives us an opportunity, at least, to look at some of the factors that may be important.

The big difference that we will have in the future is that we now have man affecting the environment in a way that he never has in the past. The problem of prediction is far more severe than the factors of environmental change in the past. I think the issue of to what extent we influence the carbon flux through runoff, producing chemical changes in our coastal waters, is something that we have very little data on at all at this stage.

HONJO: Thank you very much. We are just about coming to closing time. I want to ask for one more comment from Dr. Ishii, because he started but did not quite finish on the first deliberation about ground-truthing, which I think is very important.

ISHII: To enhance the values of a variety of observation results, I believe ground-truthing is very important. In this area also, I think what is required is international cooperation to conduct such observations.

Also, if I may put forth a request from my side if you could please give me one more minute. I discussed the next workshop just now. We have together directors of a number of different institutions, and we were able to exchange our views here. I would like to ask some other institution to host the first workshop, and our center, I think, will be ready to host the second round of workshops. So if anyone is interested in hosting the first workshop, of course our center is also ready to render support or help. I wish I had more chance to exchange opinions on this.

HONJO: Now we are running out of time, and I think we have to close the session and give the podium to President Uchida. I am supposed to give some summary of this discussion, but I feel I cannot do that because these things are so complex, with the many ideas and exciting discussions this afternoon. This is not necessarily a discussion that we can really summarize, and we anticipate having that kind of continuous discussion at the next workshop.

I would like to close the session by expressing my appreciation for your enthusiastic participation on this panel; I personally learned a lot, and I hope the floor did as well. I had the opportunity to think about this problem and really be tested on the whole of oceanography by this very large need for study on the ocean.

I happen to be the chairman of the last session of today, and that happens to be the last session of this tremendous meeting. On behalf of this panel I thank President Uchida and the members of JAMSTEC, who worked so hard, were so well-organized, kind, and sensitive. Thank you very much indeed.

MAMIYA: Thank you very much. This concludes the panel and lecture sessions of the symposium. Lastly, I would like to ask the president of JAMSTEC, Dr. Uchida, to give a summary and closing address.

ISAO UCHIDA: Distinguished guests, ladies and gentlemen, in ending this symposium on behalf of the organizer, I would like to make some remarks.

I would like to thank the heads of the institutes of oceanography of various countries for their valuable lectures and discussions. I would also like to thank the panelists and chairpersons and Dr. Kondo, who delivered the special lecture. Finally, I would like to thank the participants in this symposium for their attendance.

Through 2 days of lectures and discussions, I came to realize that each institute has its own unique history and tradition, and the activities reflect history and tradition. As institutes that specialize in ocean study, there are common appreciations as well.

Ocean research is at a turning point. At no other time in history have the social needs for oceanic research been as great as today. Researchers and specialists on oceanic study have to respond to the needs of the country, the needs of the international community, and I think we were able to share common appreciations on the direction of our future research.

First, there is the need for interdisciplinary research, organization of a number of projects, and international cooperation. There are increasing social needs for oceanic study. We have to understand the vast and complicated ocean, and we need interdisciplinary study and projects to elucidate complex stages of the ocean. Such research and study cannot be conducted by one institute alone or one country alone. International cooperation is indispensable. Especially when it comes to global environmental issues, a number of international cooperation projects and being implemented or planned and each institute is quite active in this field of study. Not only global warming, but also coastal and environmental problems are quite important in looking at the global environment. We have to prepare a database to look at that problem of environment on a global scale. We should not overreact as some of the speakers have rightly pointed out. Scientists and researchers have a mission to provide academic and scientific data. Some of the speakers also pointed out the importance of basic research in addition to project-oriented research.

The second important point is the necessity for technological development in oceanic study. The recent advances in oceanic study are supported by technological development, such as satellites, deep-sea research vessels, deep-sea drilling vessels, acoustic technology, and supercomputers. In oceanic study, synoptic simultaneous three-dimensional constant observation is going to become increasingly important. Development of technology and systems that enable such observation will more or less influence the success or failure of our research efforts.

The third important point is investment in oceanic research and development, especially investment by the government. Oceanic study is becoming extremely important, but investment in oceanic study is not commensurate to its importance. That is quite evident when we look at the budget of each institute represented here. The budget level is extremely low in comparison with the increased social needs for oceanographic study in relation to the global environment. The research is becoming larger and more sophisticated. In order to implement study, human resources and installations are required, and I think it necessary for the government to increase substantially the investment in oceanographic study.

In this symposium, we had two panel sessions on research of deep sea and ocean observations. In the panel discussion on research of the deep sea, the panelists elaborated on the possible direction of the study, the necessity of instrument development, such as manned and unmanned submersibles, drilling ships and observation stations, and the needs of the scientists were discussed during the panel session. It is my understanding that the panel was in consensus for the need for a workshop which might lead to specific joint action.

Prior to the panel session today, Dr. Kondo gave a special lecture. During the panel session, panelists shared their views on the future direction of the research and study. I shall not go into detail because that session has just been concluded. The theme is extremely important, and very little is known about the role of ocean in relation to earth's environment.

We need to understand the role of the ocean and we need international cooperation, including cooperation with developing countries. Conditions for international cooperation were taken up during the panel session, and in that sense I think that the panel was extremely important.

Based upon the fruits of the symposium, I understand that a consensus was reached to organize a workshop as a follow-up to this symposium. I hope that some institute will take the initiative in hosting a workshop; we shall appreciate such an initiative very much. As an organizer of this symposium, we would like to support such an initiative for the workshop.

That is the summary of the symposium, including my impressions of the symposium. This symposium has been represented by the head of the institutes of oceanography from a number of leading countries. The new direction of oceanographic research and development was the theme of this symposium. Our symposium has been quite successful in that we were able to identify new directions of research and development and that consensus was reached to initiate new joint actions toward new directions. I hope that cooperative relationships will be strengthened with this symposium.

Once again, I would like to thank the heads of the research institutes represented here, and I would also like to thank the audience for their attention throughout the symposium. With this, I would like to close the symposium.

Thank you very much, ladies and gentlemen.

MAMIYA: Ladies and gentlemen, thank you very much for taking part in this 2-day symposium. This concludes the International Symposium on New Directions of Oceanographic R&D.

A press conference is scheduled in Room 1102. The head of the institutes of oceanography are kindly invited to come to 1102 after this symposium. Thank you very much.

Summary and Closing Address

Summary and Closing Address

Distinguished guests, ladies and gentlemen, in ending this symposium on behalf of the organizer, I would like to make some remarks.

I would like to thank the heads of the institutes of oceanography from various countries for the valuable lectures and discussions given. I would also like to thank the panelists and chairpersons, and Dr. Kondo, who delivered the special lecture. My sincere thanks also go to other participants in this symposium for their valuable attendance.

Through the 2 days of lectures and discussions, I came to realize that each institute represented here has its own unique history and traditions which are reflected in their respective activities. At the same time, as institutes specializing in ocean study, there have been expressions of a number of common views and appreciations.

The single most important common appreciation expressed has in my opinion been that ocean research is at a turning point today. At no other time in history have societal needs for oceanographic research been so great as today, and researchers engaged in oceanographic studies have to respond not only to the needs of the country they belong to, but also to those of the international community.

I think the symposium has been able to develop a number of common appreciations as to the directions of our future research. First, there is the need for more interdisciplinary effort, enhanced organized projects, and increased international cooperation in oceanographic research. To be able to respond to the increasing social needs for oceanographic study, we have to understand the vast and complicated processes of the ocean, and we need to undertake broad and interdisciplinary studies and projects. Such research cannot possibly be conducted by any single institute or country alone. International cooperation is indispensable, especially when it comes to global environmental issues. A number of international cooperation projects are being implemented or planned, and I am impressed that each institute is already quite actively taking part in these projects in its own ways. References have been made that we will have to address not only global-scale issues such as global warming, but regional issues such as coastal environments as well. It has also been pointed out that we will have to prepare a database in this regard. We should not overreact as some of the speakers have rightly pointed out: scientists and researchers have the mission to provide scientific data to the world. Some speakers also pointed out the importance of basic research that should go side by side with project-type research.

Second, the importance of technological development in oceanographic studies has kept increasing. Recent advances in oceanic studies owe greatly to those in technology, such as satellites, deep-sea research vessels, deep-sea drilling vessels, acoustic technology, and supercomputers. In the future oceanic research and wide-area simultaneous, three-dimensional, and continuous observations will become ever more important. The development of technologies and systems that enable such observations will more or less determine the success of our research efforts.

Third, more investment in oceanographic research and development, especially investment by the government, is urgently required. Oceanic study is rapidly gaining importance; the investment in it, however, is not gaining proportionately. This is quite evident when we look at the budget of each institute represented here. The budget level is extremely low in comparison with the increased societal needs for oceanographic study addressing environment issues, global or otherwise. Research is becoming ever larger in scale and more sophisticated. In order to implement it, more human resources and equipment than currently available are required, and I think the government in each country needs to increase its investment in oceanographic study substantially.

The symposium had two panel sessions: one on deep sea research and the other on ocean observation, each constituting a major area of oceanography. In the panel session on deep sea research, the panelists discussed possible directions in future research including the necessity for developing new devices and equipment, such as manned and unmanned submersibles, drilling ships, and observation stations, and how to cope with the growing need for scientists. It is my understanding that there was a consensus among the panelists on the need for a follow-up workshop that might lead to some joint actions.

Prior to the panel session today, Dr. Kondo gave a special lecture entitled "The earth's environment and ocean research." During the panel session, the panelists exchanged their views on future directions in ocean research. I shall not go into detail because that session has just been concluded. Throughout the proceedings, however, how little we know about the ocean itself despite the extreme importance of the subject in question came home to me.

During the panel session, it was pointed out that international cooperation including the participation of developing countries would be essential to understand the complicated processes of the vast oceans. Some thoughts were also expressed on how to go about it — an important policy aspect in our future efforts, I think.

I am so glad that those panel discussions led to a consensus that workshops be organized as a follow-up to this symposium. If there are some institutes that will volunteer to host the workshops, I would of course welcome such initiatives.

However, if there are none willing to do so, and with consent from you, JAMSTEC as the organizer of this symposium would be happy to do whatever we can to arrange for the workshops.

This is my summary of the symposium, and please permit my inclusion of some personal impressions about the event. The symposium, represented by the heads of the world's leading institutes of oceanography, and conducted on the theme of "New directions of oceanographic research and development," has been quite successful, I think, in that we have been able to identify certain new directions in oceanic research and that a consensus has been reached to initiate new joint actions along the new directions identified. I would be most happy if this symposium could also lead to enhanced cooperative relationships between the participating institutions.

Once again on behalf of the organizer, I would like to thank the heads of the research institutes who came to join the event. I would also like to thank the audience for their keen attention throughout the symposium. With this, I would like to close the symposium. Thank you very much.

ISAO UCHIDA
PRESIDENT OF JAMSTEC

Keyword Index